Management in the Built Environment

The aim of this book series is to provide a platform to build and consolidate a rigorous and significant repository of academic, practice and research publications that contribute to further knowledge relating to management in the built environment. Its objectives are to:

1. Disseminate new and contemporary knowledge relating to research and practice in the built environment
2. Promote synergy across different research and practice domains in the built environment and
3. Advance cutting-edge research and best practice in the built environment

The scope of this book series is not limited to "management" issues per se because this then begs the question of what exactly are we managing in the built environment. While the primary focus is on management issues in the building and construction industry, its scope has been extended upstream to the design management phase and downstream to the post-occupancy facilities management phase. Management in the built environment also involves other closely allied disciplines in the areas of economics, environment, legal and technology. Hence, the starting point of this book series lies with project management, extends into construction and ends with facilities management. In between this spectrum, there are also other management-related issues that are allied with or relevant to the built environment. These can include, for example cost management, disaster management, contract management and management of technology.

This book series serves to engage and encourage the generation of new knowledge in these areas and to offer a publishing platform within which different strands of management in the built environment can be positioned to promote synergistic collaboration at their interfaces. This book series also provides a platform for other authors to benchmark their thoughts to identify innovative ideas that they can further build on to further advance cutting-edge research and best practice in the built environment.

If you are interested in submitting a proposal for this series, please kindly contact the Series Editor or the Publishing Editor at Springer:

Low Sui Pheng (bdglowsp@nus.edu.sg) or
Ramesh Premnath (Ramesh.premnath@springer.com)

More information about this series at http://www.springer.com/series/15765

David G. Carmichael

Future-proofing—Valuing Adaptability, Flexibility, Convertibility and Options

A Cross-Disciplinary Approach

Springer

David G. Carmichael
School of Civil and Environmental Engineering
UNSW Australia
Sydney, NSW, Australia

ISSN 2522-0047 ISSN 2522-0055 (electronic)
Management in the Built Environment
ISBN 978-981-15-0722-9 ISBN 978-981-15-0723-6 (eBook)
https://doi.org/10.1007/978-981-15-0723-6

This Springer imprint is published by the registered company Springer Nature Singapore Pte Ltd.
The registered company address is: 152 Beach Road, #21-01/04 Gateway East, Singapore 189721, Singapore

To Maria

Preface

This book is about anticipating possible futures and embracing thinking about adaptability, flexibility, convertibility and options. Possible futures contain uncertainty (implying probability, likelihood or frequency of occurrence, in contrast to determinism—Carmichael, 2014). Flexibility, adaptability and convertibility here refer to a capability to make changes in line with future circumstances. Making changes, among other things, can delay obsolescence—physical, economic, functional, technological, environmental, social or legal.

The book's presentation of a unified approach, across multiple situations and multiple disciplines, to valuing adaptability, flexibility, convertibility and options is original to the author. The book, being multidisciplinary, will be of interest to anyone involved in future-proofing, assets, investment and management.

Adaptability, flexibility and convertibility thinking fit generally within the field of options—provisions are made now that set up options to make changes or take some action in the future, but exercising these options is discretionary depending on how the future unfolds. In valuing such options, the book's approach has many strengths over existing practices.

The approach given in this book avoids all the criticisms present in the use of traditional financial option analogies to non-market-based situations. It offers a ready way to value multiple situations, requires minimal financial knowledge and mathematical sophistication, and hence can be readily implemented by practitioners. Numerical studies, to date, demonstrate that the book's approach gives acceptable estimates in valuing options. Coupled with a second-order moment analysis, the combined approach offers advantages over other methods: the calculations align with common investment and viability calculations; it is intuitively appealing and intuitive to understand; the calculations are straightforwardly performed, for example, on a spreadsheet; no assumptions are made on the probability distributions of the underlying cash flows; it does not rely on the financial market options literature including forced analogies; the concept of volatility is avoided, while acknowledging that uncertainty can exist beyond the option exercise date; there are no deterministic restrictions on any exercise price; multiple exercise dates are possible; interest rates containing uncertainty and multiple interest rates can be used; variances can differ between variables and over time; discrete or continuous time discounting is possible; and any combination of cash flows, including negative cash flows, is possible.

Financial options terminology is deliberately avoided as much as possible in this book in order to show the stand-alone nature of the book's approach. However, financial options are mentioned occasionally, in order to locate this book within the existing literature.

Sydney, Australia David G. Carmichael

Contents

About the Author

David G. Carmichael is Professor of Civil Engineering and former Head of the Department of Engineering Construction and Management at the University of New South Wales, Australia, and Distinguished Adjunct Professor at the Asian Institute of Technology, Thailand. He is a graduate of the Universities of Sydney and Canterbury; a Fellow of the Royal Society of New South Wales; a Fellow of the Institution of Engineers, Australia; a Registered Professional Engineer; a Member of the American Society of Civil Engineers; and a former graded arbitrator and mediator. Professor Carmichael publishes, teaches and consults across multiple disciplines.

Professor Carmichael is a highly respected conceptual thinker. He is regarded as a systems person, but more particularly regarded as a systems thinker coming from the orientation that engineering has given to the systems body of knowledge. Over his career, he has contributed significantly to fundamental systems thinking on engineering practices. He believes that all people can benefit from such structured and systematic thinking, but recognises that many people lack disciplined thinking and therefore may not appreciate or understand structure or system. His contributions have spanned modelling, and the fundamental systems configurations of synthesis, investigation and analysis in a range of applications. His research always has a practical application. His strength is in being able to take something, which has heretofore been presented in a complicated, esoteric and mystical way, and simplify it. Perversely, many people seem to rate highly something that they cannot understand, but quickly dismiss something presented in understandable terms.

Specific significant contributions of Prof. Carmichael, summarised in a selective bibliography at the end of the book, span the following disciplines:

- Investment—A demonstration that conventional finance theory is flawed with bad models and disguised with unnecessarily high-level mathematics and complications; a rational exposition of options analysis applicable to any real or financial option situation.
- Adaptability, flexibility and convertibility across multiple disciplines and their valuation.

- Risk—Exposing true risk and risk management in contradistinction to most people's and most publications' incorrect usage.
- Problem-Solving—A definitive treatment of problem-solving, and systems modelling, analysis, synthesis and investigation.
- Planning—A universally robust view of planning that extends to planning of any type whether project planning, strategic planning or other.
- Management and Project Management—Establishing that management, planning and design share the same structure as systems synthesis (inverse) configurations; the extension of thinking on management beyond verbal models and fads.
- Construction, Quarrying and Surface Mining Emissions—A proof of coincidence of unit costs and unit emissions within existing operations; the introduction of marginal cost abatement curves to construction; the examination of the effect of loading policies and training on emissions; a utility measure for emissions.
- Contractor Payments—An original analysis and classification of owners according to their payment characteristics.
- Contracts and Project Delivery—A systematisation and exposition of project delivery methods showing the relationship between all methods and extinguishing public discussion of risk at the delivery level; contributions to thinking on contracts—guarantees, delays and conversion.
- Construction, Quarrying and Surface Mining Operations: Production and Cost—Simplified modelling and analysis of earthmoving, quarrying and surface mining operations focusing on production and cost; extensions to the literature on queue modelling.
- Optimum Structural Design—A universally robust view of structural modelling and design; an original recognition of singularities in design.
- Material and Structure Characterisation—An original systems identification view of modelling and characterising materials and structures.

His breadth of technical knowledge is matched by few people; a polymath. He has shunned the narrow focus adopted by many academics and has published in journals and books covering multiple disciplines including structural engineering, construction, management, finance, economics, sustainability, project management, construction management, property, control systems, systems engineering and problem-solving.

Much of what Prof. Carmichael writes is considered to be left-field, and some might be considered controversial or maverick. Commonly, he shows that the status quo is flawed in an attempt to promote healthy discussion for the advancement of the state-of-the-art of the professions. The challenge for Prof. Carmichael has always been to get people to acknowledge that the state-of-the-art needs improving and not dogmatically defend current situations.

Chapter 1
Introduction

1.1 Overview

The book's approach to future-proofing and establishing the value of adaptability, flexibility, convertibility and options is original.

In a usual options scenario, a premium is paid today (that is, an up-front cost, action or equivalent) in return for having the right but not the obligation to take some action in the future. An option is a right, but not an obligation to do something. This action is referred to as exercising the option, and the option is only exercised (an action taken in the future) if it is worthwhile for the option holder to do so, and depends on future circumstances. Generally, the option would not be exercised if it was not worthwhile to do so. That is, having an option caps any downside involved at the initial cost of the premium, but rewards any upside involved. An option protects the holder against losses, no matter how large, and rewards the holder for gains, and the greater the future uncertainty the more the option is worth. This upside, when discounted to the present (time $t = 0$), allows the value of the option to be calculated. The option value may convert an initially unviable (by deterministic analysis) investment into a viable one, or make a viable (by deterministic analysis) investment more viable.

The action taken in the future can vary with the application. For example, infrastructure might be upgraded, contract terms might be changed, or something may be bought or sold (as in the financial derivatives literature). Options have broader applicability than just buying or selling and can involve many different types of management actions or decisions.

This scenario exists because of uncertainty about what will happen in the future. Uncertainty implies probability, likelihood or frequency of occurrence, in contrast to determinism [1]. Options give future flexibility. Flexibility in decision making and options go hand-in-hand. Hence flexibility has value. Having an ability to make a future choice in actions has increased value over a compulsory pre-set future action. That is, having an option increases the attractiveness of the situation.

D. G. Carmichael, *Future-proofing—Valuing Adaptability, Flexibility, Convertibility and Options*, Management in the Built Environment,
https://doi.org/10.1007/978-981-15-0723-6_1

This flexibility might be spoken of synonymously in terms such as adaptability or convertibility, depending on the application. Future-proofing for an uncertain future relies on having flexibility.

There is an allowed-for ability to adapt, convert, switch or generally take some action in the future, should it be required, but this action need not be undertaken. There is an ability (equivalent to a right) to do something in the future but not an obligation to do this. This is the core of an 'option'.

The literature on options may differentiate between real options and financial options. Financial market options (for example, in the derivatives markets) depend on the price or value of market underlyings (stock, energy, carbon, real estate, ...), while real options are based on real assets and there are no equivalent market underlyings. The approach of this book does not require that a distinction be made between real and financial options.

This scenario, involving flexibility in some future action, is present in a quite diverse range of applications. Consider some examples:

Adaptable infrastructure. Infrastructure may need future modification in order to deal with shifts in usage or demand, climate or demographics. All such issues involve uncertainty. One way of managing this is to design or build in adaptability features today in order that future change is facilitated, should it be required, but this adaptation need not be done. The infrastructure may be civil or building infrastructure.

Convertible contracts. Projects evolve over time, for example they may go through feasibility, design, construction, and commissioning and handover stages. A contract that suits one stage may not be best for another stage. This leads to having contracts that are convertible to suit the circumstances at hand at any time.

Options. Financial options involve buying and selling based on the market price or value of underlyings. Commonly this is done through stock exchanges, and are examples of derivatives. The distinction between financial and real options is historical, with financial options occurring first, and real options using financial options tools (inappropriately) by analogy. However, no analogies are necessary by the approach in this book.

Such flexibility thinking is not new, but heretofore a rational means of valuing this flexibility has not existed.

1.2 Valuation

Deterministic valuations (typically deterministic present worth) of options-style scenarios do not inform. Uncertainty needs to be accommodated in any valuation, but not through discount rates which can be shown to give rise to poor models [3]. Decision trees fail to capture future decision changes, and future decision choices. A deterministic sensitivity analysis is considered unsuitable because it says nothing about the frequency of occurrence of the sensitivity ranges of the variables; it also requires a discount rate that performs two incompatible roles—to account for the time value of money and to account for uncertainties [3]. Rather, a method acknowledging uncertainty directly is required.

Valuation methods borrowed from the financial options literature and applied to real assets are open to criticism because of the analogies used and the inappropriate modelling (such as that related to volatilities, and assumptions on time series for the prices or values of the analogous underlying) adopted. There is, as well, no mathematically-based estimation method for the prices or values of underlyings in atypical markets and atypical situations. Common methods used to value financial options are the Black-Scholes method (abbreviated to 'Black-Scholes' in the following), binomial lattices and the equivalent in Monte Carlo simulation.

By contrast, this book provides a unifying approach to valuing flexibility that covers all options-style scenarios, through a rational analysis based on probabilistic cash flows. Environmental and social issues can be incorporated. The advantages of the book's approach compared with adopting the more usual financial market analogies are outlined in Carmichael [1, 2]. Chapter 2 explores this further. The validity of the approach rests on accepting probabilistic discounted cash flow (DCF) analysis as a model for the time value of uncertain cash flows.

In summary (without the mathematics, which is given in Appendix 1.5), the value of having flexibility is:

$$\text{Flexibility or option value, } OV = \Phi M \tag{1.1}$$

where $\Phi = P[PW > 0]$ and is termed the investment feasibility [5, 6], PW is present worth, the collective values of PW > 0 are referred to as the upside, P is probability, and M is the mean of the present worth (PW) upside measured from PW = 0 (Fig. 1.1). The present worth is obtained by discounting future 'benefits', 'disbenefits' and 'costs' connected with exercising the option, but does not include the option premium. Equation (1.1) has been called the Carmichael equation. The larger the value of Φ, the more viable an investment becomes. Where $\Phi > 0.5$, there

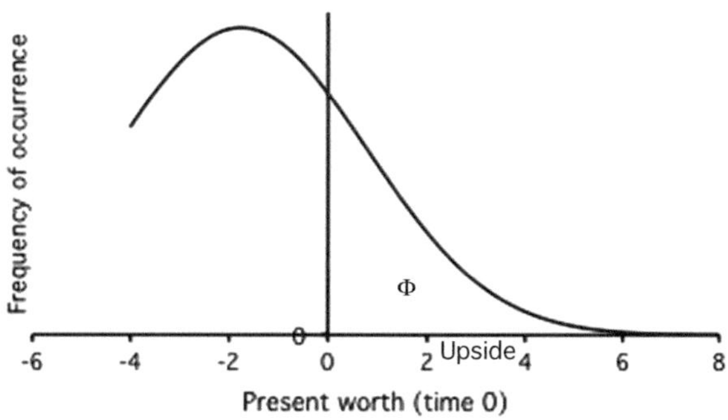

Fig. 1.1 Example upside of the present worth (PW) distribution; Φ is the area under the curve; 'mean of PW upside' is the mean of the area under the curve, both measured from the origin

may be no need for an option calculation, because deterministic calculations might already indicate viability (subject to any up-front cost or premium).

All options-style or flexibility scenarios can be valued from this. Establishing the value of this flexibility might be called options analysis in the literature.

The greater the future uncertainty, the greater is this flexibility or option value. The uncertainty in future estimates could be anticipated to increase the further into the future estimates are being made.

The value of having this flexibility is compared with the up-front premium (cost), in order to establish the viability of incorporating flexibility. For overall investment viability, OV should exceed any up-front cost or premium paid by the option holder in order to establish the option; in some cases, there is no premium paid and hence only the magnitude of OV need be looked at. With no premium, and for the particular case where the present worth expected value is positive, viability exists based on present worth alone, irrespective of any OV calculation [2].

Numerical studies [1, 2, 4] show that Eq. (1.1) is an acceptable way of valuing all options. This book's cash flow approach offers a straightforward and understandable single way of analysing all options, without recourse to the financial market options literature, avoiding most finance terminology, constraints and assumptions, while fitting within established practices of investment evaluation. It offers a ready way to value multiple options, requires minimal financial and mathematical knowledge, and hence can be readily implemented by practitioners.

1.3 Book Outline

Chapter* 2 *A Common and General Formulation

Chapter 2 gives material common to all chapters, and accordingly is referenced throughout the book. The chapter covers notation adopted, a summary of the steps to be followed in valuing any option, second order moment analysis as the book's preferred method of dealing with probabilistic discounting, estimating for financial, social and environmental variables, characterising present worth probabilistically, a comparison of the book's approach with other existing methods, and probabilistic discounting.

Chapter* 3 *Real Options

The chapter gives a cash flow view of real options in distinction from the rest of the literature on real options. Via a cash flow view, the chapter shows that real options can fit within conventional investment analysis, in particular the usual discounted cash flow analysis familiar to most practitioners, and there is no need to go looking for answers to real option valuation in other disciplines. All real option valuations can be reduced to one plain option approach or, in the case of compound options, to a collection of plain options. There is no need to distinguish option type, for example between expand or contract. The chapter will be of interest to anyone involved in investment in real assets.

Chapter* 4 *Adaptable Infrastructure—Civil

For longevity, infrastructure (civil and building) needs to cope with any shift in social and environmental states, and any shift in technology that may occur in the future. Having infrastructure, with built-in flexibility, capable of addressing these shifts and prolonging the point in time at which infrastructure becomes obsolete, would therefore appear to be sensible, provided that it can be shown to be viable, when compared with the more usual non-built-in flexibility case. The viability analysis necessarily requires the inclusion of the uncertainty surrounding future shifts, their magnitude, timing and costing. Heretofore, estimating the value of built-in flexibility has been done deterministically or qualitatively, or by analogy with methods derived from the financial options literature, even though it is generally acknowledged that there are shortcomings in such approaches. By contrast, this chapter gives a general probabilistic treatment that overcomes these shortcomings and is generally applicable across all infrastructure types and across all change types. The uncertainty in costing as well as the economic uncertainty in intangibles related to social and environmental matters are addressed and incorporated into the calculations. With greater uncertainty, it is shown that the inclusion of flexibility becomes more viable, and flexibility leads to improved sustainability performance. Viability is examined through a comparative analysis of the built-in flexibility case versus the non-built-in flexibility case. The chapter gives illustration examples on road infrastructure. The chapter will be of interest to those designing, building and managing infrastructure, as well as those with an interest in the sustainability of infrastructure.

Chapter* 5 *Adaptable Buildings and Houses

Requirements of building (including housing) owners commonly transition over time. Organisations, individuals and families transition over time, for example as markets shift, individuals age and families grow and decrease in size. It would thus seem appropriate that buildings should be built with adaptability features that facilitate change. There are financial, social and environmental impacts associated with altering and moving buildings and houses. With possible future alteration in mind, this chapter looks at the viability of deliberately incorporating flexibility into buildings at the time they are designed and built, as compared with no specifically incorporated flexibility (yet still possibly capable of being altered). However, although adaptability is incorporated it may not be called upon, depending on future circumstances. A comparative analysis, rather than an absolute analysis, is outlined. As case examples, the chapter considers the viability of incorporating deliberate two-storey flexibility into a single-storey house in order to provide increased future space, and dividing a house into two smaller separable house spaces, where smaller future space is desired. It can be shown through examples that incorporating deliberate built-in flexibility performs positively against all sustainability criteria—financial, social and environmental, separately or combined—however the generality of this conclusion remains to be proven. The chapter will be of interest to those at the design stage of buildings, including houses.

Chapter* 6 *Project Delivery—PPP Guarantees

The chapter is written for Public-Private Partnership (PPP, P3) toll road projects, but it applies generally to all PPP type projects. Recent publications have shown the role that options play in addressing the uncertainty, risk and fairness in public-private partnership (PPP, P3) agreements. Typically, the literature uses financial market options techniques, and applies these by analogy; each option is presented and analysed in stand-alone papers. In contrast, this chapter presents the book's original single unifying approach for analysing all PPP toll road options. The chapter will be of interest to anyone involved in PPP projects.

Chapter* 7 *Project Delivery—PPP Concession Periods

The chapter is written for Public-Private Partnership (PPP, P3) toll road projects, but it applies generally to all PPP type projects. A defining characteristic of a PPP is a concession period. A concession period that is too long financially benefits the concessionaire in the later years, at the expense of the travelling public—the public continues to pay tolls for longer times than it should. A concession period that is too short does not allow the concessionaire to fully recoup its initial investment or provide a suitable return for the investment. The issue becomes—what is a reasonable concession period? The chapter suggests a way forward, providing flexibility based on actual project performance, with an option available to the authority to take over the operation of the road. The chapter's approach eliminates the controversial aspects associated with concession periods. As a by-product, the approach additionally eliminates another controversial matter, namely that surrounding the setting of tolls—toll formulae and toll adjustments over time.

Chapter* 8 *Project Delivery—CDM

The chapter puts forward a proposal that, within Clean Development Mechanism (CDM) projects, investors be allowed to benefit from options; this will require a CDM rule change. Through the presence of options, the downside risk resulting from low carbon prices and/or low achieved emission reductions on projects can be limited, while any upside resulting from high carbon prices and/or high achieved emission reductions can be taken advantage of. It is demonstrated that the presence of options improves the financial attractiveness of CDM projects, and this is at no detriment to any stakeholder. The flow-on from the proposal is that more CDM projects should be realisable if options are available, and this in turn will lead to reduced global emissions and improved sustainability. The proposal is supported by the necessary theory and is demonstrated on two CDM projects, one on hydropower and one on wind power.

Chapter* 9 *Convertible Contracts

Commonly, projects start out with broadly defined information and this gets refined as the project progresses. This suggests that a prudent approach would be to tailor the contract between the project owner and the project contractor to the respective project stage, with conversions along the way. Information asymmetry between owner and contractor also suggests the need to tailor a contract to a project's situation. In line with this, the chapter explores the role and viability of changing contract terms as a project progresses, utilising the book's original approach to convertibility. Discussion is given on owner and contractor risk, the common law issues associated with implementing such conversions, any compensation that the owner might need to pay, the timing of the conversion, and associated practical implementation issues. The chapter, for definiteness, concentrates on construction contracts with conversion between payment types, but the chapter's approach applies to all contracts and all terms within contracts. Having convertibility within a contract can be shown to offer benefits to both contracting parties. The idea of having flexible contracts is not new, but heretofore a rational method of analysing their value has been missing. Current literature does not deal directly with the matter addressed in the chapter.

Chapter* 10 *Financial Options

The chapter presents a method for estimating the value of a financial option using the book's probabilistic present worth analysis. The method is shown to capture the upside value of a financial option in an equivalent way, and give similar results, to the Black-Scholes method. The strength of the book's approach is shown to lie in its intuitive appeal, the avoidance of having to estimate volatility, relaxed assumptions, and the simplicity of the calculations. The likeness with the Black-Scholes method is given, with differences noted, and confirmed numerically for a range of input parameters. The approach given in this book is applicable to evaluating any option.

Chapter* 11 *Energy Options

The energy price in deregulated markets exhibits uncertainty, volatility, spikes and seasonality characteristics. Hedging with energy options (energy derivatives) provides one way that players in the market deal with the price uncertainty and associated risk. Common financial option pricing tools, such as modified Black-Scholes and binomial and trinomial lattices, can be used to value the options. However, in atypical markets, including those where something unforeseen happens in energy production or supply, or there are shifts in economic conditions or external factors, such pricing tools may not apply. This chapter shows how energy options can be valued in such atypical markets as well as regular markets using the book's approach. Different types of energy options are explored—plain call and put options, spark spread options and swing options, and options associated with callable and putable forwards.

Chapter* 12 *Real Estate Options

Options occur in many places within real estate or property transactions. Hedging through options provides one way of dealing with uncertainties and associated risk. Commonly, option pricing techniques such as Black-Scholes, binomial lattices or equivalent Monte Carlo simulation might be used. However, in atypical markets where property fluctuations are anticipated to not follow any usual patterns, such pricing techniques may not apply. External factors, particularly related to shifting economic and political environments, may lead to future behaviour differing from the past. This chapter demonstrates how all property-related options can be valued in such atypical markets as well as regular markets. Various property-related options are outlined in terms of how they can be valued by the book's approach.

1.4 Future Directions

The book outlines a cash flow approach to evaluating all options. Future research could examine many possible refinements, for example in the choice of the present worth distribution and assumed interest rates, and in examining special-case options.

Like other methods, such as Black-Scholes, over time users of the book's approach will come to understand the approach's strengths and weaknesses and what fine adjustments need to be made in order suit particular adaptability, flexibility, convertibility and option requirements.

The applications of future-proofing are very broad. The book covers many interesting applications but not all. Some suggested applications that readers may like to pursue relate to soft systems such as people with their career changes and shifting skill requirements, and organisations with over- and under-staffing to deal with uncertain future demand.

1.5 Appendix: The Carmichael Equation—Derivation

The Black-Scholes equation is a commonly accepted method of evaluating European financial options. (A 'European' option is exercised on a specified date; an 'American' option can be exercised at any time before or on the specified date.) Some underlying has a price or value, movement in this price or value is assumed to follow geometric Brownian motion with a constant volatility over time, and prices or values are assumed to be lognormally distributed; a constant risk-free interest rate is used. Users of the Black-Scholes approach adjust the standard Black-Scholes' assumptions to suit their applications. For definiteness, let the underlying be some stock (but in other applications it could be energy, carbon, or other), and it is the stock price which is the focus.

The Black-Scholes equation (call or buy option) can be reinterpreted to provide directly an estimate of the value of an option using familiar present worth analysis and assumptions less-restrictive than that required for Black-Scholes.

The following refers to the exercise price (denoted K) and the stock price (denoted S) as cash flows. From the buyer's (option holder's) viewpoint, for a call option, the exercise price is regarded as a negative cash flow, while the stock price is regarded as a positive cash flow. Consider exercising at time T. At time T, a stock price S_T greater than K reflects the option value or upside.

Define the option value at time T, OV_T, as the expected value of the net cash flow at T, evaluated on the assumption that the option is only exercised if it is worthwhile (in the money, $S_T > K$). Then,

$$OV_T = E^*[\text{Net cash flow}_T] = E^*[S_T - K]$$

E[] denotes expected value. To denote that the option is only exercised if it is worthwhile (in the money), the symbol * is used. OV_T can now be discounted to time 0. Using continuous time discounting, in order to compare with Black-Scholes, an estimate of the option value at time 0 with an interest rate r is,

$$\begin{aligned} OV &= e^{-rT} OV_T = E^*[e^{-rT} S_T - e^{-rT} K] \\ &= \text{Expected}^* \text{ PW of stock price} \\ &\quad - \text{Expected}^* \text{ PW of exercise price} \end{aligned}$$

OV is evaluated assuming that the option is only exercised if it is worthwhile (in the money). In effect, OV is being calculated as in,

$$OV = e^{-rT} E[\max(S_T - K, 0)]$$

For a lognormal assumption on the stock price and a risk-free rate, this leads to the Black-Scholes equation; a proof of this is given for example in Hull [7].

That is, the call option value in Black-Scholes, and OV above are obtained through expressions that are structurally the same. In both cases, the assumption is that the option will only be exercised if it is worthwhile.

A distinction between the exercise price and the stock price was made above in order to make a comparison with Black-Scholes and for no other reason. However in terms of getting to OV, there is no need to make this distinction; both the stock price and the exercise price are treated as cash flows. Accordingly, assuming that the option is only exercised if it is worthwhile, that is there is no exercising (probability of zero) if the present worth is negative,

$$\begin{aligned} OV &= E^*[PW] \\ &= (1 - \Phi) \times 0 + \Phi \times M \\ &= \Phi M \end{aligned}$$

where PW is the present worth of all cash flows at time 0, M is the mean of the present worth upside, and Φ is referred to as feasibility and equals $P[PW > 0]$, the upside area in Fig. 1.1 [1, 2, 4]. The mean of the present worth upside is measured from $PW = 0$. The situation at time 0 is shown in Fig. 1.1. The term 'upside' is used here to denote positive present worth, and represents a worthwhile deal.

All options, adaptability, flexibility and convertibility can be valued by the single Eq. (1.1). For example, a put (sell) option is treated no differently to a call (buy) option. Each option calculation is interpreted from the option holder's viewpoint.

Everything which is favourable to the option holder is regarded as a cash inflow, and everything unfavourable is regarded as a cash outflow; strict accounting conventions need not be used. Any cash flows connected with exercising an option are discounted to give PW, and hence OV from Eq. (1.1). The discounting of the cash flows, which generally are random variables, can be done through a second order moment analysis as in this book (or the equivalent numerically-oriented Monte Carlo simulation). A second order moment analysis requires no assumptions to be made on the distributions of the cash flows, but rather works in terms of expected values, E[], and variances, Var[], of the cash flows. Possible ways of estimating the expected values and variances are discussed in the Chap. 2, along with a second order moment analysis.

Higher uncertainty in the stock price leads to greater spread in the present worth distribution, and hence higher option values, as in the Black-Scholes method.

Feasibility, Φ

Feasibility, Φ, is a measure that establishes the suitability of an investment. Where competing investment choices exist, that with the largest feasibility might be preferred. With an individual investment, the question arises as to what is a level of feasibility acceptable to the investor, that is, what is an acceptable level of probability that the present worth will turn out to be positive. The answer to this will depend on whether the investor is risk prone, risk averse or risk neutral, and hence requires knowledge of the investor's attitude. A small value of Φ (close to 0) might be considered equivalent to being far from or deep out of the money; a large value of Φ (close to 1) to being far from or deep in the money; while a value of Φ of approximately 0.5 to being close to the money.

OV is used as an estimate of the option value. It is the upside potential of the investment. Where competing investment choices exist, that with the largest option value might be preferred (dependent on the cost/price/premium of the option right). With an individual investment, what is considered a minimum acceptable option value will depend on other circumstances and intangibles surrounding the investment, and the cost/price/premium of the option right.

As drawn in Fig. 1.1, the mean or expected value of the total present worth distribution is less than 0, implying that the deterministic present worth of the investment is less than zero, and hence under conventional deterministic thinking, the investment would not be undertaken.

It is noted that there will always be an option value because of the positive tail of the distribution representing PW. However the option value will approach zero for investments with low feasibility, Φ. And so investors might prefer to make a decision based on both the option value, OV, in conjunction with the feasibility, Φ.

In the probabilistic case, that is where uncertainties exist, investments may turn out as losses and also turn out as gains, with probabilities attached. In the deterministic case, the situation is more black and white—the present worth is either positive or non-positive.

Put-call parity

Put-call parity can be shown to hold [1].

American option

For an American option, let the time to exercising the option follow a probability distribution. Let p_t be the probability that the exercise takes place in period t, t = 0, 1, 2, …, T, and let $PW_{(t)}$ be the associated present worth. Then,

$$E[PW] = \sum_{t=0}^{T} p_t E[PW_{(t)}]$$

$$Var[PW] = E[PW^2] - \{E[PW]\}^2$$

The choice of distribution for the exercise time is at the discretion of the person doing the calculations. Example possible distributions are uniform where exercising is equally likely at any time up to T; a declining triangular distribution where the value of the option is perceived as decreasing with time; or a rising triangular distribution where exercising is more likely the closer time gets to T. The case $p_T = 1$ and $p_t = 0, t = 0, 1, 2, \ldots, T - 1$ is the European option.

References

1. Carmichael DG (2014) Infrastructure investment: an engineering perspective. CRC Press, Taylor and Francis, London
2. Carmichael DG (2016) A cash flow view of real options. Eng Econ 61(4):265–288
3. Carmichael DG (2017) Adjustments within discount rates to cater for uncertainty—guidelines. Eng Econ 62(4):322–335, 2017
4. Carmichael DG, Hersh AM, Parasu P (2011) Real options estimate using probabilistic present worth analysis. Eng Econ 56(4):295–320
5. Carmichael DG, Balatbat MCA (2008) Probabilistic DCF analysis, and capital budgeting and investment—a survey. Eng Econ 53(1):84–102
6. Carmichael DG, Balatbat MCA (2008) The influence of extra projects on overall investment feasibility. J Financ Manag Propert Construc 13(3):161–175
7. Hull JC (2006) Options, futures and other derivatives, 6th ed., Prentice Hall, Upper Saddle River, New Jersey

Chapter 2
A Common and General Formulation

2.1 Notation

The main notation, common to all chapters, is listed here. Notation specific to applications is outlined within the relevant chapter.

Cov[]	covariance
COV	coefficient of variation; StDev/Mean
E[]	expected value
i	time (typically, years) counter; i = 0, 1, 2, …, T, …, n
M	mean of the PW upside, measured from the origin
n	life of the investment
OV	option value estimate obtained from Eq. (1.1)
P	probability
PW	present worth obtained by discounting the X_i typically to i = 0; refer Appendices 2.11.1–2.11.3
r	interest rate; a subscript i denotes an interest rate in year i.; r can be probabilistic and is then characterized by its expected value $E[r_i]$ and variance $Var[r_i]$.
StDev[]	standard deviation, $\sqrt{Var}$
T	time of exercising the option if at one point in time; exercising is also possible over an extended time period
Var[]	variance; standard deviation squared
X_i	the net cash flow in period i, i = 0, 1, 2, …, n $X_i = a_{i1}Y_{i1} + a_{i2}Y_{i2} + \ldots + a_{im}Y_{im}$, where a_{ik} are constants (typically +1 and −1). X_i is characterized by its expected value $E[X_i]$ and variance $Var[X_i]$. There may be correlation between the net cash flows; expressions for the expected value and variance of X_i are given in Appendices 2.11.2 and 2.11.3.

D. G. Carmichael, *Future-proofing—Valuing Adaptability, Flexibility, Convertibility and Options*, Management in the Built Environment,
https://doi.org/10.1007/978-981-15-0723-6_2

Y_{ik} cash flow component k in period i, i = 0, 1, 2, …, n; k = 1, 2, …, m; Y_{ik} is characterized by its expected value $E[Y_{ik}]$ and variance $Var[Y_{ik}]$; the cash flow components can be both revenue and cost related, or positive and negative, or cash inflows or outflows; there may be correlation between the cash flow components. The magnitudes of the variances (and expected values) can be different for different cash flows and can be different over time if desired.

ρ correlation

Φ feasibility, $P[PW > 0]$

Terminology

The book covers multiple flexibility situations across different disciplines, and establishing the valuing of this flexibility. Generally, this implies an 'investment' and an 'investor' who might gain from having this flexibility. Accordingly, the term 'investment' is used in a generalised sense. The term 'investor' may interchange with the terms 'user' or 'option holder'. The investor or option holder is taken as the person doing the calculations according to the book's approach. The value of an option is calculated in this book from the option holder's viewpoint. The term 'user' is in the sense of using the book's approach to value options.

The term 'option' is also generally used, from which particular examples of adaptability, flexibility and convertibility arise.

2.2 Outline

The unified approach, given in this book, to valuing adaptability, flexibility, convertibility and options across all disciplines is obtained by regarding any investment involving one or more options as a collection of cash flows (cash inflows and cash outflows), which are regarded as random variables. The cash flows directly connected with exercising an option are isolated (from all the cash flows). These cash flows are discounted, typically to a (probabilistic) present worth and, in interpreting the collective present worth distribution, the option value is calculated from Eq. (1.1). The strengths of the book's approach are that it requires minimal mathematical background and sophistication, is straightforward and intuitive to understand and implement, and is a natural extension of and hence has the familiarity of existing (deterministic) investment appraisal practices; the book's approach reduces to conventional deterministic present worth analysis in the absence of uncertainty.

The cash flows (cash inflows and cash outflows) associated with exercising an option are looked at from the point of view of whoever is holding or exercising the option. For time of exercising denoted T, such cash flows may occur at T only, or at and beyond T; cash flows beyond T can be discounted to T in order to give a more familiar options look of having something that looks like cash flows at T only. The term 'cash flow' is used broadly, and not just that which appears in business accounts, and can include for example something lost because of exercising the option, direct and indirect costs, prices, and income foregone. An option is only exercised generally if, at the time, the anticipated cash inflows (at and beyond T) exceed the anticipated cash outflows (at and beyond T). No distinction needs to be made as to the option type (for example expand or contract, or buy or sell, as highlighted in the options literature) because the signs of the cash flows are solely considered from the point of view of the investor who holds the option (to exercise or not). This is consistent, for example, with usual feasibility studies [3]. Conventional accounting standards are not used. Where exercising an option alters existing cash flows, it is this difference in cash flows (between that without exercising and that with exercising), which is relevant.

Cash flows, which are not connected with the option, are not included in the calculations (but are obviously part of the greater investment viability analysis).

In order to have an option available or to create an option at some later point in time, there typically is some up-front cost or premium. This cost is not included in the option calculation, but rather the calculated option value is compared with this up-front cost, in order to establish the viability of creating the option. This up-front cost could be anticipated to be more where the available options are multiple, and/or a choice as to which option to exercise is available.

An interest rate, with no loading for cash flow uncertainty in the investment, is used, because uncertainty is already being taken care of through the cash flows, but also possibly through the interest rate. However, any appropriate interest rate can be adopted in the calculations, different rates can be used for different cash flows if desired, and probabilistic rates can be used. This is consistent with usual feasibility thinking, and is not constrained by traditional financial options assumptions. Different interest rates for different periods of time, different interest rates for different sources of money, and different interest rates for cash inflows compared with cash outflows, can be adopted. The book's approach recognizes this, and allows users of the book's approach to adopt whatever rates they feel are most applicable.

Where multiple options are present, the same situation as for a single option repeats according to how the multiple options interact. Some discounting, however, may then be to a nominated time different to the present day, for example to a time where one project stage transitions to another project stage.

In establishing when is the best time to exercise an option, different available times of exercising, T, can be considered in the analysis and option values compared via enumeration, leading to the best T value.

The book's approach avoids traditional financial market option pricing tool assumptions, analogies and formulae. The book's approach offers a single way to calculate the value of any option in any discipline in any situation.

2.3 Summary Option Calculation Steps

The following is a summary of the steps involved in calculating any option value, and is used throughout this book:

1. Obtain estimates for the expected values and variances of any cash flows (inflows and outflows) or equivalent cash flows directly resulting from exercising an option. Equivalently, obtain estimates for the expected values $E[Y_{ik}]$ and variances $Var[Y_{ik}]$ of the cash flow components Y_{ik}, $i = T, T + 1, T + 2, \ldots, n$; $k = 1, 2, \ldots, m$, in period i.
2. The nature of these cash flows or equivalent cash flows could be anticipated to be different with each application.
3. Calculate the expected value of the present worth, E[PW], and the variance of the present worth, Var[PW], by discounting collectively these cash flows. See Appendices 2.11.2 and 2.11.3.
4. Fit a probability distribution to the present worth, PW. The choice of distribution is discretionary.
5. Calculate Φ and M.
6. Calculate the option value, OV, from Eq. (1.1).

Each of these steps—estimating, discounting, distribution fitting, and calculating Φ, M and OV—is discussed in the following sections and appendices of this chapter. Expected values and variances are spoken of (rather than probability distributions) because of the book's preferred use of second order moment analysis.

2.4 Second Order Moment Analysis

In getting to the distribution of the present worth, any method preferred by the person doing the calculations can be used. The author's preference (and the one adopted in this book) is to use a second order moment analysis [1–3] leading to an expected value and variance of the present worth, to which a distribution can be fitted.

A second order moment analysis is adopted because of the closed-form nature of the approach, and the considerable insight that it gives. Monte Carlo simulation (but not doing equivalently to Black-Scholes) is an alternative, but being a numerical method it provides little insight while also requiring assumptions on the probability distributions for each of the independent variables (cash flows and interest rates), and such distributions would rarely be available.

For a second order moment analysis, variables are characterised in terms of their expected values, E[], and variances, Var[], referred to as moments. These may vary over time. Uncertainty is embodied in the variance terms. Where a variable is deterministic, its variance is zero. Expressions can be derived quite readily connecting the moments of the dependent variable with the moments of the independent variables. Any covariance or correlation between variables has to be acknowledged. Then, a two-parameter probability distribution, for example a normal distribution, can be readily fitted to the dependent variable; for example, Ang and Tang [1] lists the parameters of common distributions and their relationships to expected values and variances.

Appendices 2.11.2 and 2.11.3 give the relationships between the moments of present worth, PW, and the moments of the cash flows and interest rates.

2.5 Estimating Cash Flows

2.5.1 *Moments*

Estimates of expected values and variances of cash inflows and cash outflows may be done in any reasonable way through usual estimating practices. In the absence of anything else, the following is a useful approach. Firstly, optimistic (a), most likely (b) and pessimistic (c) values are estimated as is done in the Program Evaluation and Review Technique (PERT) [18]. This then gives, expected value or mean = $(a + 4b + c)/6$, and variance = $[(c - a)/6]^2$ [4, 6] Distributions other than that used in PERT, such as uniform, might be preferred. Other methods are given in Carmichael [3]. For example, a proxy approach to estimating variance can be used if the investor has previous similar projects, by analysing the variances of their similar cash flows or cash flow components. Or investors may prefer to estimate maximum, minimum and most likely values, use a triangular distribution, and calculate an expected value and variance based on this. Persons doing the calculations are free to adopt whatever methods they like in order to estimate the characteristics of future cash flows.

It is remarked that estimating is not an exact or precise science. Estimates might be adjusted based on: historical data or trends, experience, education, 'gut feel' (a combination of logic and emotion), personal knowledge, the market, national and international economies and events; and/or subjective probability estimates; and/or based on current news and forecasts of the future.

Trends might be based on time series developed from historical data, and adjusted for outliers and seasonality; an estimate might then be called a forecast or prediction. However, time series are not so useful in atypical markets, where the past does not carry forward to the future.

For stock or other market-based entity, estimates might be based on: a time series of historical stock prices; using a proxy approach if the investor is aware of similar stocks; or if the stock volatility has been estimated, then the variance of the stock price can be obtained from this through Appendix 2.11.4, and the expected value of the stock price from the future worth of the current stock price.

The future contains uncertainty. Uncertainty in the estimates could be anticipated to increase with time; data will be less well known into the future. Such increases in uncertainty over time can be accommodated by estimating larger variances with time [3].

This approach to estimating is consistent with conventional investment appraisal thinking, and is not constrained by any financial theory assumptions; estimating is always a mixture of knowledge, education, experience and anything else which is relevant, and no two estimators will arrive at the same estimate values. Such estimates are done to a level of accuracy comparable to other estimates in the particular discipline.

2.5.2 Correlations

Correlations between the cash flows may also be needed, and these might best be estimated based on knowledge of the particular components and their paired relationships, rather than through any involved mathematical analysis.

Existing data would generally not be available on the covariances or correlations of cash flows, or if it was available, accurate values could not be anticipated. Correlation is discussed in Kim and Elsaid [15], Kim et al. [16], Johar et al. [14] and Carmichael [3], among others. Estimates of correlations may be done in any reasonable way. At best, only approximate estimates may be available. The two cases of (i) statistical independence, and (ii) perfect correlation, bound actual correlations, and provide an envelope to the actual correlation. To estimate the correlation coefficient between two cash flows, the only way forward may be to reason logically, using physical arguments, as to what the relationship is between the two. And then supplementing this with experience, expertise, and knowledge of the situation to finally establish a reasonable estimate. If data are available, using inbuilt correlation coefficient calculations in spreadsheets might be helpful [3].

2.6 Estimating Disbenefits and Benefits

2.6.1 General

In situations that not only involve cash flows, but also social and environmental intangibles, the literature might prefer to use the terminology of costs, disbenefits and benefits. Costs are inputs to any investment and are naturally measured in dollars, while disbenefits and benefits are outputs to the investment and have various units of measurement, one of which is dollars [3]. Disbenefits and benefits might be handled in several possible ways:

i. All social and environmental disbenefits and benefits are converted to equivalent dollars, enabling them to be treated similarly to costs, and the analysis becomes one of being purely financial. The conversion to monetary equivalents may be through life cycle assessment (LCA) or social life cycle assessment (SLCA) and economics practices, though it is acknowledged that this is not without controversy.

ii. All social and environmental disbenefits and benefits are left in their original units of measurement and considered in parallel with the financial analysis. Commonly, feasibility studies focus on the financial issues but also include some form of narrative on intangibles.

iii. Combining financial, social and environmental issues into one, perhaps using weighting functions to reflect the importance and units of measurement of the social and environmental issues.

In later chapters, costs, disbenefits and benefits are dealt with in a number of ways, dependant on the situation. For example, where a comparison analysis is done, it might be assumed that the benefits of the alternatives are the same, and so benefits are omitted from the comparison calculation.

Environmental issues involved in, for example, adapting infrastructure include: resource consumption (materials and energy); emissions (equipment-produced emissions; and embodied emissions in materials); and waste generation (waste and reuse). Rating scales, as used in assessing the environmental credentials of infrastructure, are not suitable for adaptability analysis purposes, but rather preference is for life cycle assessment (LCA) style calculations at time $t = 0$ (initial construction) and $t = T$ (adaptation), perhaps converted to monetary values.

The treatment of social issues can be done similarly to environmental issues, namely through social life cycle assessment (SLCA) style calculations at $t = 0$ and $t = T$, converted to monetary values. Social issues, for example in adapting infrastructure, result from the physical and/or psychological impacts on stakeholders, including that connected with construction work. Some relevant social issues include those associated with: workers (health and safety, and employment); the public (traffic, pollutants, dust, noise and vibration); and owners (disruption and inconvenience).

Care has to be exercised that double-counting does not occur between financial costs, and social and environmental disbenefits, and correlation between costs and disbenefits has to be considered as being possible.

2.6.2 Social and Environmental Estimates

Placing a monetary value on intangibles is not without controversy.

Some social and environmental issues, where markets exist, can be directly estimated. For example, carbon emissions might be converted to a monetary value through the use of traded carbon credits or a carbon tax, if available. Quantifying intangibles and placing a monetary value on intangibles, where no market exists, is not a precise art, but can provide decision makers with the information they need to make high level decisions. It is this imprecision in the estimates which needs to be captured in the variance estimates. This imprecision is exampled below. Gilchrist and Allouche [11], among others, outline various techniques for quantifying social 'costs'. Generally, the analysis is done from the viewpoint of those impacted and, where multiple groups are impacted, some form of mediated consensus is required.

Typical disbenefits that might need inclusion in any analysis are accidents, waste, noise, dust, vibration, pollutants, emissions, temporary relocation, traffic disruption, and travel time.

How each of these might be treated can be exampled on the issue of waste. There are many possible ways that waste might be valued such as: (1) the actual cost of disposal; (2) the sale value; (3) the cost to extract raw materials, produce and transport an equivalent product; (4) the cost of recycling minus the sale price of the refurbished product; (5) the cost of alternative reuse minus the benefit of this; (6) any legislated fines; (7) loss of future business due to poor sustainability or corporate social responsibility credentials; (8) the trade-off between increased production with waste, and decreased production without waste; and (9) future captured methane as an energy source. It is conceivable that not everyone will agree on the same method of valuation, and it is also conceivable that multiple methods of valuation could have some applicability.

Each of the z valuation methods, available for intangible κ, will give a value $Z_{\kappa\varphi}$, $\varphi = 1, 2, \ldots z$. Depending on each method's relevance, applicability and credibility in the opinion of the analyst, each method can be given a weighting $w_{\kappa\varphi}$. A value for $w_{\kappa\varphi}$ of zero indicates that the method is not relevant or not applicable to the situation at hand, or is based on poor assumptions. Weights normalized so as to sum to 1 would be preferred. Techniques such as the analytic hierarchy process (AHP) can be used here, or the weightings can be based on opinion alone. Alternatively, where a valuation method itself gives some indication of its accuracy, and on the basis that a method with lower dispersion in its estimate is possibly more accurate, weights might be chosen in proportion to the inverse of a method's calculated variance, such as $w_{\kappa\varphi} = \Lambda_{\kappa\varphi} / \sum_{\varphi} \Lambda_{\kappa\varphi}$ where $\Lambda_{\kappa\varphi}$ is the inverse of $\mathrm{Var}[Z_{\kappa\varphi}]$.

Appropriate moments can be estimated in various ways, but in the absence of anything else, the following ways might be adopted, in conjunction with the optimistic-most likely-pessimistic approach mentioned earlier, and conventional statistical methods of calculating sample means and sample variances. Feasibility study level of accuracy is implied, not detailed estimate level of accuracy. For cash flow Y_κ,

$$E[Y_\kappa] = \sum_{\varphi=1}^{z} w_{\kappa\varphi} E[Z_{\kappa\varphi}]$$

$$\mathrm{Var}[Y_\kappa] = \sum_{\varphi=1}^{z} w_{\kappa\varphi}^2 \mathrm{Var}[Z_{\kappa\varphi}] + 2 \sum_{\varphi=1}^{z-1} \sum_{\eta=\varphi+1}^{z} w_{\kappa\varphi} w_{\kappa\eta} \mathrm{Cov}[Z_{\kappa\varphi}, Z_{\kappa\eta}]$$

Here the last term disappears when the methods' valuations are independent. It is again remarked that estimating is not a precise science, and depends on the estimator's experience, education, access to data, and available modelling.

2.7 Discounting

For typical projects, discrete time yearly, monthly or daily discounting, using a yearly, monthly or daily interest rate respectively, might be favoured over continuous time discounting, because of its familiarity and ready incorporation into spreadsheets. Both discrete time and continuous time discounting will lead to similar conclusions, but discrete time discounting is more in tune with the way practitioners think.

2.8 Distribution for Present Worth

Persons doing the calculations are free to assume or choose whatever distribution for present worth that they think is appropriate. Knowing E[PW] and Var[PW], derived through Appendices 2.11.2 and 2.11.3, any probability distribution can be fitted to PW, but it is anticipated that most people would use a normal distribution [3, 5, 12, 19].

Where the present worth is the sum of many components (each with uncertainty, and each discounted), then the distribution of present worth could be anticipated to follow a normal distribution based on the central limit theorem [12].

A normal distribution for PW is assumed in the book's numerical calculations, but any other suitable distribution could be used.

The shape of a normal distribution is completely defined on knowing its expected value and variance, and associated probabilities are readily evaluated using standard normal probability tables.

The equation for the probability density function of a normal distribution for a random variable Q is,

$$f_Q(q) = \frac{1}{\sqrt{2\pi}\sigma}\exp\left[-\frac{1}{2}\left(\frac{q-\mu}{\sigma}\right)^2\right] \qquad -\infty < q < \infty$$

with parameters μ and σ. These are related to the expected value and variance as follows,

$$E[Q] = \mu \qquad Var[Q] = \sigma^2$$

These may be used to find the particular shape of the normal distribution in any circumstance (that is, using the so-called 'method of moments'). (Note that this is not the same σ as in Appendix 2.11.4).

2.9 Calculating Φ and M

In Fig. 1.1, Φ is the area under the curve (PW upside), while M is the mean value of the part distribution shown, measured from the origin.

Φ and M are readily established numerically using a spreadsheet, among other ways.

For any assumed probability distribution for present worth, for approximate values, the upside part of the present worth distribution can be divided into vertical strips and its area and centroid calculated as a structural engineer would calculate for a member cross section. For strip s, s = 1, 2, …, S, of width Δ, height h_s (obtained by evaluating the probability density function),

$$\Phi = \text{PW upside area} = \sum_{s=1}^{S} h_s \Delta$$

$$M = \text{Mean of PW upside} = \frac{\sum_{s=1}^{S} h_s \Delta}{S\Delta}$$

The number of strips used will be determined by whatever accuracy is desired.

Alternatively, denoting Q as present worth, and $f_Q(q)$ as the present worth distribution,

$$\Phi = \int_0^{\infty} q f_Q(q) dq$$

$$M = \frac{\int_0^{\infty} q f_Q(q) dq}{\int_0^{\infty} f_Q(q) dq}$$

2.10 Comparison with Financial Market Options Techniques

For option calculations that are able to be done using methods such as the Black-Scholes method (abbreviated to 'Black-Scholes'), the results using the book's approach are almost the same; normalized with respect to an equivalent exercise cost, all comparisons lie within a few per cent. Accuracies much less than this occur in the input cash flow estimates in most investments, while the presence of intangibles decreases the accuracy further. However, the book's approach is capable of dealing with far more situations and far less restrictions than those assumed within Black-Scholes and other financial options analysis methods. Users of the book's approach are also able to incorporate their individual knowledge and perceptions in addition to any other information, qualitative or quantitative, that is available or can be forecast. Structurally, Black-Scholes and the book's approach capture the value of an option in equivalent ways. Numerical results in Carmichael et al. [10] and Carmichael [3, 5] confirm similar results. Exact correspondence with Black-Scholes is not anticipated because a direct one-for-one substitution of variables or one-for-one transformation between variables is not possible, in particular volatility-to-variance and time series-no time series.

As to which gives the better options values or estimates—financial options methods or the book's approach—cannot be stated because the two are not directly comparable. However it is believed that, for real options, the assumptions behind the book's approach are more relevant, and hence the options values or estimates are more likely to be better.

Carmichael et al. [10] and Carmichael [3, 5] outline the most obvious differences with, and advantages over, Black-Scholes. In terms of the book's approach:

- An underlying's price or value is not required to follow any time series (with associated volatilities), including particular series such as geometric Brownian motion. The book's approach is thus applicable where the future is not anticipated to repeat the past, while also being applicable where the future is anticipated to repeat.
- No assumptions are made on the distribution of an underlying's price or value at any time; rather, with a second order moment approach, cash flows are characterised by their expected values and variances, and the present worth distribution chosen is discretionary.
- The exercise cost/price need not be deterministic and a single value; Eq. (1.1) allows for general exercise prices containing uncertainty and not necessarily a single value at one point in time.
- Uncertainty is incorporated through variance terms of the cash flows (and possibly interest rates), and not through any volatility measure; uncertainty can vary throughout time and is not restricted to an equivalent constant volatility, and can be different for different cash flows. Uncertainties attached to cash flows are at the discretion of the person doing the calculations. No volatility estimate is required. It is noted that there is no consensus in the existing literature on how to determine the value of volatility for real options—"There is no single, theoretically justified approach for calculating the volatility coefficient for real options" [17]. Volatility is inapplicable and not transferrable to most real options situations. Real options formats cannot be made to fit financial options formats exactly unless simplifications and assumptions are made. The book's approach can capture uncertainty applying not only at the time of exercising but also beyond that time.
- Interest rates can be chosen at the discretion of the person doing the calculations, can be probabilistic, can vary throughout time, and can be different for different cash flows and different for positive and negative cash flows.
- Where exercising an option can alter subsequent cash flows, this can be accommodated in the book's approach but not by Black-Scholes. Various cash flows, and cash flow correlations, in any combination, can be accommodated.

- Discounting may be either continuous time or discrete time.
- The book's approach applies to both assets that are not traded and assets which are traded.
- There is no need to distinguish between different option types; each option and associated cash flow signs are established relative to the option holder.
- The book's approach is intuitive and an extension of existing discounted cash flow (DCF) analysis that is commonly used by industry. No mathematical sophistication is needed in order to understand the book's approach.

For real options, the book's approach avoids traditional financial market option pricing tool analogies, is directly applicable, and also is consistent with conventional feasibility study practice. Issues involved in estimating analogous volatilities are avoided in favour of using cash flow variances. The book's approach thus has considerably wider application than traditional financial option analysis methods.

2.11 Appendices

The following appendices are presented from elementary to more complicated. Appendix 2.11.1 gives the deterministic expressions for present worth, from which the expressions in Appendices 2.11.2 and 2.11.3 follow. Appendix 2.11.2 gives expressions for the expected value and variance of present worth, PW, deriving from cash flows only as random variables. Appendix 2.11.3 extends this to allow both cash flows and interest rate to be random variables. A further extension to allow for the occurrence of cash flows at probabilistic times is possible, and follows Carmichael and Balatbat [7], but is not repeated here.

Strict accounting conventions are not followed. Rather, everything is looked at from the viewpoint of the option holder, such that a cash flow favourable to the holder is taken as positive and a cash flow unfavourable to the holder is taken as negative. The option is valued from the viewpoint of the option holder.

Appendix 2.11.4 gives the conversion used in numerical examples in this book, between variance and volatility.

2.11.1 Appendix: Deterministic Expressions

Consider a general investment, with possible cash flows extending over the life, n, of the investment. Let the net cash flow at each time period, $i = 0, 1, 2, \ldots, n$, be the result of a number of cash flow components (random variables), $k = 1, 2, \ldots, m$. The cash flow components can be both revenue (or equivalent) and cost (or equivalent) related. There may be correlation between the cash flow components at the same period.

The net cash flow X_i in any period can be expressed as,

$$X_i = a_{i1}Y_{i1} + a_{i2}Y_{i2} + \ldots + a_{im}Y_{im}$$

where Y_{ik}, i = 0, 1, 2, …, n; k = 1, 2, …, m, is the cash flow in period i of component k, and a_{ik} are constants (either +1, 0 or −1). Or using expected value notation, E[],

$$E[X_i] = \sum_{k=1}^{m} a_{ik}E[Y_{ik}] \tag{2.1}$$

The present worth, PW, is the sum of the discounted X_i, i = 0, 1, 2, …, n, according to,

$$PW = \sum_{i=0}^{n} \frac{b_i X_i}{(1 + r_i)^i} = \sum_{i=0}^{n} PW_i$$

where r_i is the interest rate in year i, and b_i are constants (either +1, 0 or −1), and PW_i is the present worth due to X_i. Or using expected value notation,

$$E[PW] = \sum_{i=0}^{n} \frac{b_i E[X_i]}{(1 + E[r_i])^i} \tag{2.2}$$

The interest rate, r_i, if constant drops the subscript i, and different rates can be attached to different cash flows if desired.

Continuous time discounting can be used if preferred. Variances do not exist because the variables are deterministic.

2.11.2 Appendix: Probabilistic Cash Flows

Denoting E[] and Var[] as expected value and variance respectively,

$$E[X_i] = \sum_{k=1}^{m} a_{ik}E[Y_{ik}] \tag{2.3}$$

$$Var[X_i] = \sum_{k=1}^{m} a_{ik}^2 Var[Y_{ik}] + 2\sum_{k=1}^{m-1}\sum_{\ell=k+1}^{m} a_{ik}a_{i\ell}Cov[Y_{ik}, Y_{i\ell}]$$

Alternatively, the variance expression can be written in terms of the component correlation coefficients, $\rho_{k\ell}$, between Y_{ik} and $Y_{i\ell}$, k, ℓ = 1, 2, . . . , m,

$$Var[X_i] = \sum_{k=1}^{m} a_{ik}^2 Var[Y_{ik}] + 2\sum_{k=1}^{m-1}\sum_{\ell=k+1}^{m} a_{ik}a_{i\ell}\rho_{k\ell}\sqrt{Var[Y_{ik}]}\sqrt{Var[Y_{i\ell}]} \tag{2.4}$$

For independence between Y_{ik} and $Y_{i\ell}$,

$$\mathrm{Var}[X_i] = \sum_{k=1}^{m} a_{ik}^2 \mathrm{Var}[Y_{ik}]$$

For Y_{ik} and $Y_{i\ell}$ perfectly correlated,

$$\mathrm{Var}[X_i] = \left(\sum_{k=1}^{m} a_{ik} \sqrt{\mathrm{Var}[Y_{ik}]} \right)^2$$

The expected value and variance of the present worth become,

$$E[PW] = \sum_{i=0}^{n} \frac{b_i E[X_i]}{(1 + r_i)^i} \tag{2.5}$$

$$\mathrm{Var}[PW] = \sum_{i=0}^{n} \frac{b_i^2 \mathrm{Var}[X_i]}{(1 + r_i)^{2i}} + 2 \sum_{i=0}^{n-1} \sum_{j=i+1}^{n} \frac{b_i b_j \mathrm{Cov}[X_i, X_j]}{(1 + r_i)^{i+j}}$$

Alternatively, the variance expression can be written in terms of the intertemporal correlation coefficients between X_i and X_j, namely ρ_{ij}, rather than the covariance of X_i and X_j,

$$\mathrm{Var}[PW] = \sum_{i=0}^{n} \frac{b_i^2 \mathrm{Var}[X_i]}{(1 + r_i)^{2i}} + 2 \sum_{i=0}^{n-1} \sum_{j=i+1}^{n} \frac{b_i b_j \rho_{ij} \sqrt{\mathrm{Var}[X_i]} \sqrt{\mathrm{Var}[X_j]}}{(1 + r_i)^{i+j}} \tag{2.6}$$

For independent cash flows X_i,

$$\mathrm{Var}[PW] = \sum_{i=0}^{n} \frac{b_i^2 \mathrm{Var}[X_i]}{(1 + r_i)^{2i}}$$

For perfect correlation of the cash flows X_i,

$$\mathrm{Var}[PW] = \left(\sum_{i=0}^{n} \frac{b_i \sqrt{\mathrm{Var}[X_i]}}{(1 + r_i)^i} \right)^2$$

Var[PW] is smaller for the assumption of independence compared with the assumption of perfect correlation. Correlation of cash flows will produce a larger present worth variance and larger option value.

The interest rate, r_i, if a constant drops the subscript i, and different rates can be attached to different cash flows if desired. Variances can differ with cash flows and over time if desired. Continuous time discounting can be used if preferred.

2.11.3 Appendix: Probabilistic Cash Flows and Interest Rates

Appendix 2.11.2 allows for uncertainty only in cash flows and can be extended and made more general. This appendix gives expressions for the expected value and variance of present worth, PW, deriving from both cash flows and interest rate as random variables [8, 9].

Expressions for $E[X_i]$ and $Var[X_i]$—Eqs. (2.3) and (2.4)—still apply. Expressions for E[PW] and Var[PW] become more general. In terms of PW_i,

$$E[PW] = \sum_{i=0}^{n} E[PW_i]$$

$$Var[PW] = \sum_{i=0}^{n} Var[PW_i] + 2\sum_{i=0}^{n-1}\sum_{j=i+1}^{n} Cov[PW_i, PW_j]$$

Alternatively, the variance expression can be written in terms of the intertemporal correlation coefficients, ρ_{ij}, between PW_i and PW_j,

$$Var[PW] = \sum_{i=0}^{n} Var[PW_i] + 2\sum_{i=0}^{n-1}\sum_{j=i+1}^{n} \rho_{ij}\sqrt{Var[PW_i]}\sqrt{Var[PW_j]}$$

This can be simplified for both the independent and correlated PW_i cases.

$E[PW_i]$, $Var[PW_i]$ and $Cov[PW_i, PW_j]$ can be established through examining,

$$PW_i = \frac{X_i}{(1 + r_i)^i}$$

and using first order and second order approximations based on a Taylor series. Such approximations are acceptable provided the coefficients of variation are not large and the original function (here PW_i) is not too nonlinear [1, 2]. Both conditions are satisfied here. For X_i and r_i uncorrelated random variables,

$$E[PW_i] = \frac{E[X_i]}{(1 + E[r_i])^i} + \frac{i(i+1)E[X_i]}{2(1 + E[r_i])^{i+2}} Var[r_i]$$

This equation shows how the value of a general investment depends on interest rate variance. For increased interest rate variance, the expected value of PW_i (and hence the expected value of PW, and feasibility Φ) increases for net positive cash flow, but decreases for net negative cash flow.

That is, for general investments, for a given net positive cash flow, an increase in the variance of the interest rate increases the present worth of the investment. This influence of interest rate variance is increased with increasing life of the investment and increasing size of expected cash flows. More investments could be anticipated to become viable with increasing uncertainty in the interest rate. For a given net negative cash flow, the situation reverses.

Where $E[X_i]$ is negative, then the present worth will not only decrease directly due to these negative $E[X_i]$ (the first term), but it will decrease further due to the variance (second) term.

Continuing, and still with X_i and r_i uncorrelated,

$$\mathrm{Var}[\mathrm{PW}_i] = \frac{1}{(1+E[r_i])^{2i}}\mathrm{Var}[X_i] + \left(\frac{i^2E^2[X_i]}{(1+E[r_i])^{2i+2}}\right)\mathrm{Var}[r_i]$$

This equation shows how the present worth variance depends on interest rate variance. For increased interest rate variance, the variance of PW_i (and hence the variance of PW) increases, irrespective of whether the net cash flow is positive or negative.

That is, for general investments, for given cash flows, an increase in the variance of the interest rate increases the variance in the present worth. This influence of interest rate variance is increased with increasing life of the investment and increasing size of expected cash flows.

Continuous time discounting could be used if preferred. Interest rates and variances can be selected as being different for different cash flows and at different points in time if required.

2.11.4 Appendix: Variance and Volatility

In places, the book compares the results using the book's cash flow approach with that of Black-Scholes. This is not possible everywhere because of the restrictive assumptions behind Black-Scholes. The result given in Hull [13] is adopted in this book to guide the conversion between variance and volatility, while acknowledging that a single formula for converting between variance and volatility has not been agreed upon by researchers [17],

$$\sigma = \sqrt{\frac{\ln\left(\frac{\mathrm{Var}[\Theta_T]}{E[\Theta_T]^2}+1\right)}{T}}$$

Here, σ is volatility, Θ is the variable being considered, and T is the time at which the conversion is being done.

References

1. Ang AH-S, Tang WH (1975) Probability concepts in engineering planning and design, vol I. Wiley, New York
2. Benjamin JR, Cornell CA (1970) Probability, statistics, and decision for civil engineers. McGraw-Hill, New York
3. Carmichael DG (2014) Infrastructure investment: an engineering perspective. CRC Press, Taylor and Francis, London
4. Carmichael DG (2006) Project planning, and control. Taylor and Francis, London
5. Carmichael DG (2016) A cash flow view of real options. Eng Econ 61(4):265–288
6. Carmichael DG, Balatbat MCA (2008) Probabilistic DCF analysis and capital budgeting and investment—a survey. Eng Econ 53(1):84–102
7. Carmichael DG, Balatbat MCA (2011) Risk associated with managed investment primary production projects. Int. Project Organ Manag 3(3/4):273–289
8. Carmichael DG, Bustamante BL (2014) Interest rate uncertainty and investment value—a second order moment approach. Int J Eng Manag Econ 4(2):176–189
9. Carmichael DG, Handford LB (2015) A note on equivalent fixed-rate and variable-rate loans. Borrower's Perspect Eng Econ 60(2):155–162
10. Carmichael DG, Hersh AM, Parasu P (2011) Real options estimate using probabilistic present worth analysis. Eng Econ 56(4):295–320
11. Gilchrist A, Allouche EN (2005) Quantification of social costs associated with construction projects: state-of-the-art review. Tunn Undergr Space Technol 20(1):89–104
12. Hillier FS (1963) The derivation of probabilistic information for the evaluation of risky investments. Manage Sci 9(3):443–457
13. Hull JC (2002) Options, futures and other derivatives, 5th edn. Prentice Hall, New Jersey
14. Johar K, Carmichael DG, Balatbat MCA (2010) A study of correlation aspects in probabilistic NPV analysis. Eng Econ 55(2):181–199
15. Kim SH, Elsaid HH (1988) Estimation of periodic standard deviations under the pert and derivation of probabilistic information. J Bus Financ Account 15(4):557–571
16. Kim SH, Hussein HE, Kim DJ (1999) Derivation of an intertemporal correlation coefficient model based on cash flow components and probabilistic evaluation of a project's NPV. Eng Econ 44(3):276–294
17. Lewis NA, Eschenbach TG, Hartman JC (2008) Can we capture the value of option volatility? Eng Econ 53(3):230–258
18. Malcolm DG, Roseboom JH, Clark CE (1959) application of a technique for research and development program evaluation. Oper Res, 646-669
19. Tung YK (1992) Probability distribution for benefit/cost ratio and net benefit. J Water Resourc Plann Manag 118(2):133–150

Chapter 3
Real Options

3.1 Introduction

A real options analysis values future flexibility in a real asset context. The flexibility relates to having a future choice between alternatives, or having the future ability to influence the directions of a project or venture, for example to contract, abandon or expand a project in order to improve envisaged outcomes. Real options have wide relevance throughout industry, including applications in resources, research and development, patents, contracts and adaptation to shifting climates and demographics. Examples in this book emphasize the diverse range of applications.

Historically, the published literature on real options adopted established financial options methods, such as the Black-Scholes equation (Black-Scholes) and binomial lattices and equivalent calculations using Monte Carlo simulation (as distinct from Monte Carlo simulation as an alternative to a second order moment analysis as mentioned in Chap. 2), and applied them analogously to real assets. This is understandable because financial options analysis existed before real options were formalized. Unfortunately, many assumptions applying to financial options do not translate well to real options.

Using the book's cash flow approach and Eq. (1.1), this chapter covers a complete collection of plain and compound real options—contract, abandon, choice, switch, delay/deferment, sequential, parallel, and rainbow options. A common treatment of all real options is given. This is demonstrated on examples. It is shown that there is no need to distinguish between the different real option types; each is reduced to its respective cash flows, and analysed using the approach outlined in Chaps. 1 and 2.

The literature on real options is very large, but its usage has been inhibited by its adoption of financial options assumptions and modelling, with an associated high level mathematical requirement.

D. G. Carmichael, *Future-proofing—Valuing Adaptability, Flexibility, Convertibility and Options*, Management in the Built Environment,
https://doi.org/10.1007/978-981-15-0723-6_3

Real options share some of the characteristics of financial options, but also have some important differences. Chief among the differences are:

- Financial options rely on the movement in the price of a market underlying and this movement is largely not able to be influenced by the investor, whereas the cash flows in a real option can derive from and be dependent on influenceable sources including management of the real asset.
- Real options and real assets are not traded, are specific to an organization, are not proprietary in nature, and generally have no market comparables.
- Traditional financial options analysis is based on the price of a market underlying described by a time series with volatility. For most real options there is no equivalent market underlying (but rather, there is a real asset). Volatility may have no transferable meaning for non-traded assets.
- Financial options are embedded in contracts specifically defining exercise prices and expiry dates, and exercising rules are clearly defined, whereas real options commonly are embedded within management decisions with discretionary parameters.
- The exercising of financial options can bring about an instant return, whereas exercising and the benefits of exercising a real option may occur over a long period.

For these reasons, traditional methods of analysis, developed for financial options, are not directly applicable to real options and have been criticized. These differences with financial options, together with the analogy assumptions (real equivalents to underlyings, exercise price and volatility), lead to much of the criticism of adopting traditional financial option pricing tools for real options, even though generally the value of a real option is dependent on similar influences to those that determine the value of a financial option. But having said that, many people are comfortable using traditional financial options methods to analyse real options situations.

A number of writers have pointed out the deficiencies in using traditional financial option pricing methods for real options [7]. A main deficiency centres on the treatment of volatility of the asset price or value, with no real agreement as to how this should be done, and the assumption that it remains constant over time. Other deficiencies relate to deterministic exercise prices, the assumption of geometric Brownian motion or similar, a market place for the asset, lognormal assumptions, exercising instantaneously rather than over a period of time, known limits on the time of exercising rather than times that might be poorly defined, continuous time discounting, the zero-sum outcome, and exercising not affecting the asset value [7].

The chapter outlines a common treatment of all real options. This is demonstrated on examples. Because, among other things, all the cash flows in the following are random variables, financial options methods are not applicable. However, for the expand and contract example calculations, comparisons with Black-Scholes values are given, based on restricting the assumptions in the examples. Numerical comparisons in the examples are given as the difference in the calculated option values, normalized with respect to the exercise cost (or equivalent); the volatility-variance conversion given in Appendix 2.11.4 is adopted for these comparisons. Other numerical comparisons are given in Carmichael et al. [7] and Carmichael [5, 6].

3.2 Real Option Types

Real option types might be classified as:

- Single or plain (stand-alone) options (expand, contract, abandon, switch, delay/defer, rainbow); or
- Compound or combined (multiple plain) options (sequential, parallel, choice).

Compound options may have embedded independent and dependent options. An option is independent if its value can be calculated separately from what is happening with other options. An independent option may affect another option's cash flows or prevent the exercising of another, and may be a pre-requisite or co-requisite for another option. The value of a collection of independent exercisable options is the sum of the individual option values. Dependence between any of the options in this collection will give a total value different to this.

The cash flows and option values for each option type are discussed here. It is seen that there is no need for a classification distinguishing between plain option types, because the same thinking applies to all.

The notation follows that given in Chap. 2. The cash flow component k in period i, is denoted Y_{ik}, $i = 0, 1, 2, \ldots, n$; $k = 1, 2, \ldots, m$. These get discounted to the present worth, PW, following Appendix 2.11.2. A normal distribution is assumed for PW in calculating the option value, OV. Interest rate, r, is set at 10% for the examples.

3.3 Option to Expand

A typical expand scenario is where (future) additional capital investment leads to an enlarged operation, perhaps to exploit a new or growing market, and generating additional cash inflow. Examples include upgrading infrastructure, follow-on development of a manufacturing project and investment in research and development.

The relevant cash flows to use in the option calculation, for a greenfield-style expansion, are the cost of expansion (negative) together with any cash flows resulting from the expansion (usually both positive and negative). Where an expansion affects existing cash flows (brownfield), the relevant cash flows to be used in the analysis are the cost of expansion together with the difference between what would have existed (assuming no expansion) and what new cash flows result from the expansion (usually both positive and negative).

Example Consider a possible expansion at year 3. The expansion cost is estimated as: $E[Y_{32}] = -\$90k$ and $Var[Y_{32}] = (\$12k)^2$. This will generate both positive and negative cash flows. The positive cash flows are estimated as: $E[Y_{i1}] = \$20k$ and $Var[Y_{i1}] = (\$5k)^2$; the negative cash flows are estimated as: $E[Y_{i2}] = -\$4k$ and $Var[Y_{i2}] = (\$1k)^2$, i = 4, 5, …, 10 (Fig. 3.1). Y_{i1} and Y_{i2}, i = 4, 5, …, 10 are assumed independent, while the X_i, i = 3, 4, …, 10 are assumed well correlated.

Discounting, $E[PW] = -\$9.10k$, $Var[PW] = (\$9.64k)^2$. The upside of the present worth distribution is given in Fig. 3.2. From Eq. (1.1), OV = \$0.89k.

(Black–Scholes cannot deal with this example directly. To compare with Black-Scholes, a deterministic expansion cost (strike or exercise price) is required—for this situation, \$90k at year 3. Consistent with established real options practice, the asset value at year 3 is obtained from the net cash flows after year 3 discounted to year 3. The asset value at year 0 is the asset value at year 3, discounted to year 0. Asset value variance is converted to volatility as indicated above. Black-Scholes gives an option value approximately 1.5% lower, as a proportion of the exercise price.)

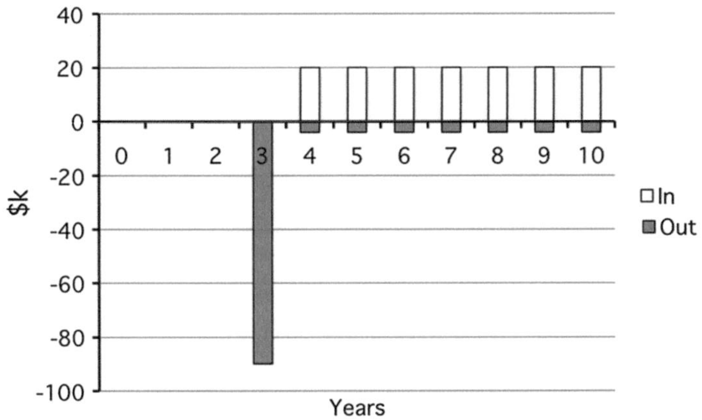

Fig. 3.1 Example—option to expand; expected values of cash flows

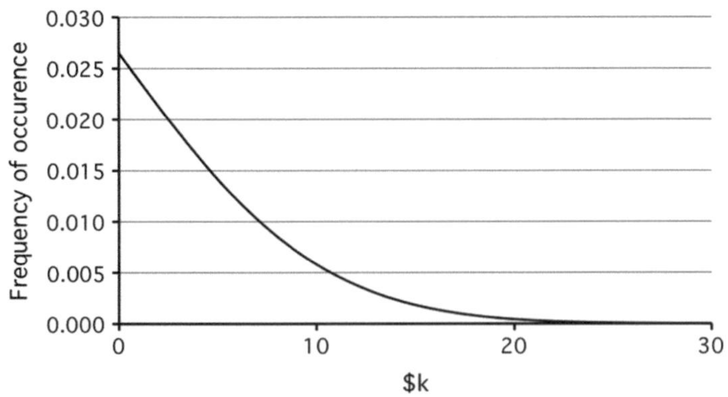

Fig. 3.2 Upside of PW distribution; expand example

3.4 Option to Contract

An example contract scenario involves an organization cutting back on operations, downsizing, or selling part of the organization (in the future), as a result of an anticipated or real downturn in demand. It could involve the income from the sale of part of an asset, or the cost involved with disposal of part of an asset. The option is exercised if the benefits from contracting outweigh those from continuing the status quo. The option to contract can protect an organization from losses in the future resulting from the effects of an unfavourable market. Contraction might allow resumption to former or greater production levels through expansion further into the future, when the market improves.

The relevant cash flows to use in the option calculation are the difference between what would have existed (assuming no contraction) and what new cash flows result from the contraction (usually both positive and negative), together with any sale income or disposal costs.

Example An option to contract, for example through a reduction in operations at a future time, may result in benefits foregone. Consider a possible sale of part of an operation at year 2. The sale price is estimated as: $E[Y_{21}] = \$71k$ and $Var[Y_{21}] = (\$5k)^2$. Without the sale, the anticipated net positive cash flows of the operation have expected values and variances, respectively, of \$27k and $(\$3k)^2$, giving: $E[Y_{i2}] = (-)\$27k$ and $Var[Y_{i2}] = (\$3k)^2$, i = 3, 4, …, 7. With the sale, the net positive cash flows are anticipated to reduce to: $E[Y_{i1}] = \$8k$ and $Var[Y_{i1}] = (\$1k)^2$, i = 3, 4, …, 7 (Fig. 3.3). Y_{i1} and Y_{i2}, i = 3, 4, …, 7, and the X_i, i = 2, 3, …, 7 are assumed well correlated.

The option to contract is dealt with in the same way as the option to expand. From the investor's viewpoint the cash flows might be thought of as being reversed.

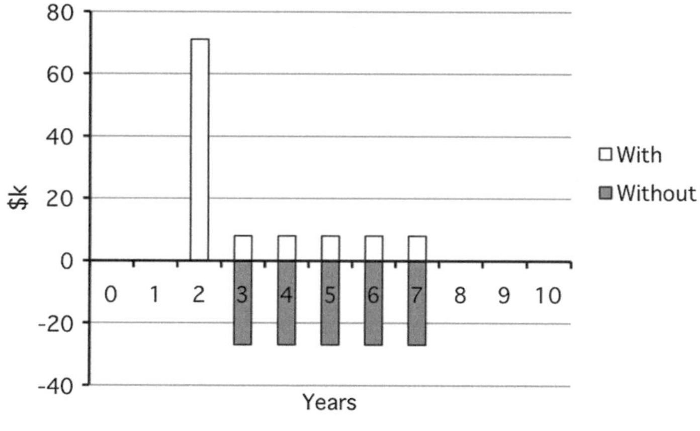

Fig. 3.3 Example—option to contract; expected values of cash flows

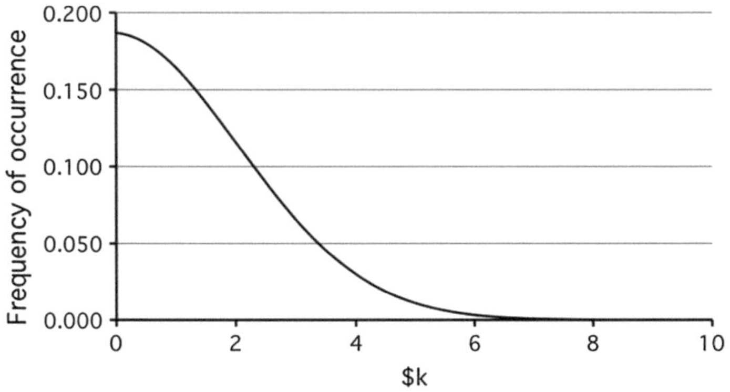

Fig. 3.4 Upside of PW distribution; contract example

Discounting, E[PW] = −$0.85k, Var[PW] = ($2.13k)^2. The upside of the present worth distribution is given in Fig. 3.4. From Eq. (1.1), OV = $0.80k.

(Black–Scholes cannot deal with this example directly. To compare with Black-Scholes, a deterministic sale price (strike or exercise price) is required—for this situation, $71k at year 2. Consistent with established real options practice, the asset value at year 2 is obtained from the net cash flows after year 2, discounted to year 2. The asset value at year 0 is the asset value at year 2, discounted to year 0. Asset value variance is converted to volatility as indicated above. Black-Scholes gives an option value approximately 1% lower, as a proportion of the exercise price.)

3.5 Option to Abandon

The option to abandon might be regarded as a particular case of contraction (or contraction might be regarded as partial abandonment). At abandonment, there may be an additional one-off cash flow—the salvage cost (negative, cash outflow) or residual (positive, cash inflow) value. The option is exercised if the savings made from abandonment outweigh future revenues that would have existed without abandonment. An example of abandonment is the early termination of a research and development project due to anticipated failure should the research continue. Abandonment might occur at any stage of any project.

Example The option to abandon is analysed as a particular case of contraction, or contraction might be regarded as partial abandonment. The cash flows are similar in nature for contraction and abandonment. For example, consider the case analysed in Sect. 3.4. If this was abandonment rather than contraction, then $E[Y_{i2}]$ and $Var[Y_{i2}]$, i = 3, 4, …, 7, are all zero (Fig. 3.5).

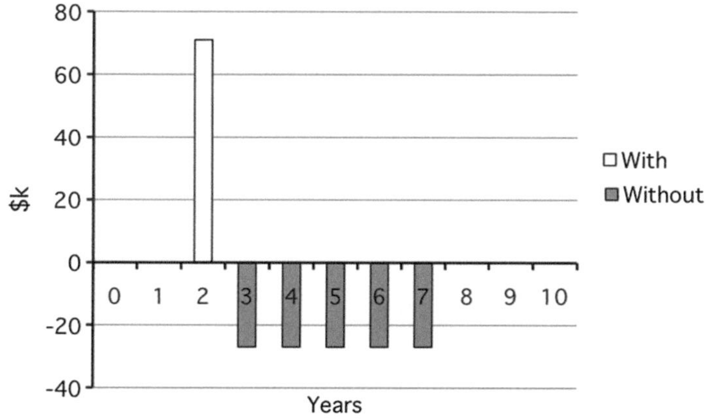

Fig. 3.5 Example—option to abandon; expected values of cash flows

Sell-back agreement. Consider acquiring an asset accompanied by a sell-back option after one year. Two possible option situations or cases relate to: (1) the sell-back price is pre-agreed (certain, or deterministic); and (2) the sell-back price is uncertain and depends on the market. Case 1 closely mirrors a traditional financial options framework, and might be dealt with analogously using Black-Scholes. However Case 2 does not fit a traditional financial options framework.

To demonstrate numerically, for Case 1, let the sell-back price be $E[Y_{11}] = \$1M$, while for Case 2, assume that this is the (market-based) estimated sell-back price expected value with a (market-based) estimated sell-back price variance of $Var[Y_{11}] = (\$0.1M)^2$. Let the estimated expected value and variance of the asset value at the end of year 1 be $E[Y_{12}] = (-)\$1.1M$, and $Var[Y_{12}] = (\$0.05M)^2$ respectively. The cash flow diagram, for both cases, from the viewpoint of the holder of the sell-back option is shown in Fig. 3.6.

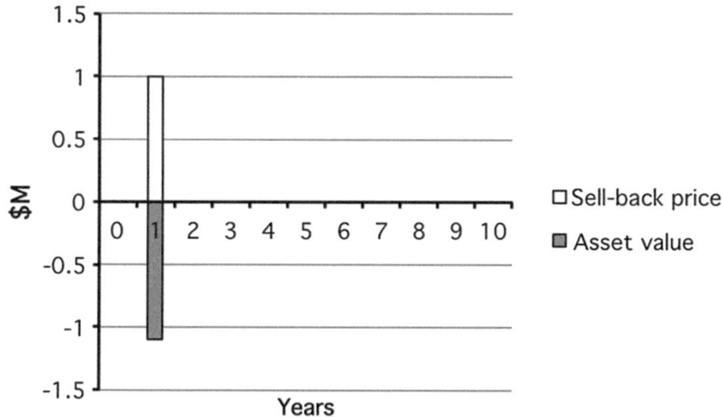

Fig. 3.6 Cash flows—sell-back example

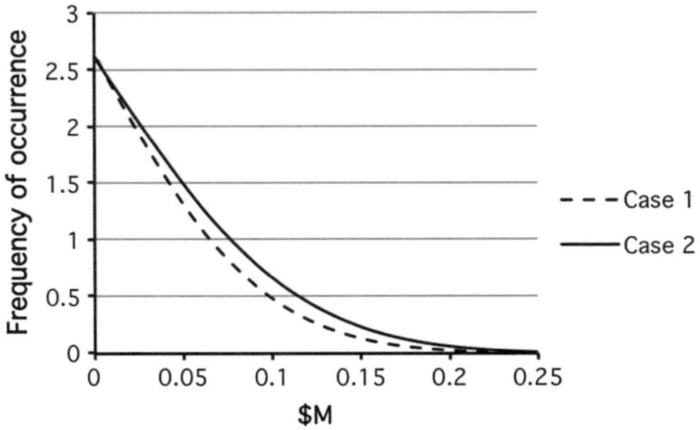

Fig. 3.7 Upsides of PW distributions—sell-back example

Discounting Y_{11} and Y_{12} gives E[PW] = –$0.091M for both cases, and Var[PW] of ($0.091M)^2 and ($0.102M)^2 respectively for Case 1 and Case 2. This gives an option value, OV, of $0.0076M and $0.0103M respectively for Case 1 and Case 2.

(For Case 1, Black-Scholes gives an option value approximately 2.5% different, as a proportion of the sell-back price. Generally, it has been found that agreement with Black-Scholes is better with call-style options rather than put-style options, as in this example. This is believed to be due mainly to the differing distribution assumptions between Black-Scholes and this book's approach (the values given here are based on a normal distribution for PW). Assuming a different PW distribution, in conjunction with the book's approach, could give closer agreement with Black-Scholes.)

Figure 3.7 plots the PW upsides, showing the difference between the two cases. The option value for Case 2 is higher because of the increased uncertainty, which is reflected in the higher variance of PW

3.6 Choice in Option Types

At a future point in time, there may be, say, the possibility either to expand, to contract, to abandon or to continue operations unchanged. For example, dependent on ore prices, yield and reserves, a mining company might expand or contract operations, abandon the mine, or continue operations unchanged. In this scenario, only the most attractive option is exercised. That is, each option is valued separately and the option yielding the highest value, at the future time, is exercised if it is better than continuing operations unchanged. Each option is mutually exclusive because it is not possible to exercise more than one option at a given time; for example, an organization cannot simultaneously expand and contract a project. Also, exercising one of the possible options alters the subsequent investment cash flows, and creates

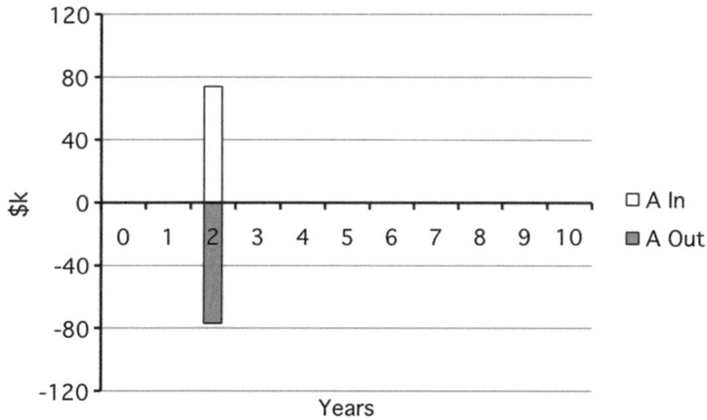

Fig. 3.8 Example—expand option A at year 2; expected values of cash flows

the need to recalculate the values of the other options based on the new cash flows. In principle, the value of having a choice in options should be higher than or equal to the value of any of the component options, because of the enhanced flexibility.

Care needs to be exercised when comparing options that are capable of being exercised at different times, or have cash flows of different orders of magnitude, much like the situations in conventional investment preference calculations using deterministic present worth.

Example Consider the possibility of selecting between two expand options, capable of being exercised from year 2 onwards. Only one option would be exercised, not both. For the first option (A), the expansion cost discounted to year 2 is: $E[Y_{22}] = -\$77k$, $Var[Y_{22}] = (\$5k)^2$, while the return discounted to year 2 is: $E[Y_{21}] = \$74k$, $Var[Y_{21}] = (\$7k)^2$ (Fig. 3.8). For exercising in later years, the standard deviation of the return is assumed to grow by 12.5% per year, with other values staying constant. For the second option (B), the expansion cost discounted to year 2 is: $E[Y_{22}] = -\$75k$, $Var[Y_{22}] = (\$5k)^2$, while the return discounted to year 2 is: $E[Y_{21}] = \$74k$, $Var[Y_{21}] = (\$12k)^2$. For exercising in later years, the expansion cost is assumed to grow by 1% per year, with other values staying constant. Within each option, the expansion cost and return are assumed to be well correlated.

Discounting to give E[PW] and Var[PW], and using Eq. (1.1) for OV, Fig. 3.9 shows option value with time, indicating a crossover of preference of one option over the other. The value of having a choice follows the upper envelope represented by the preferred option at any time. If either option A or B is exercised, this alters any subsequent analysis.

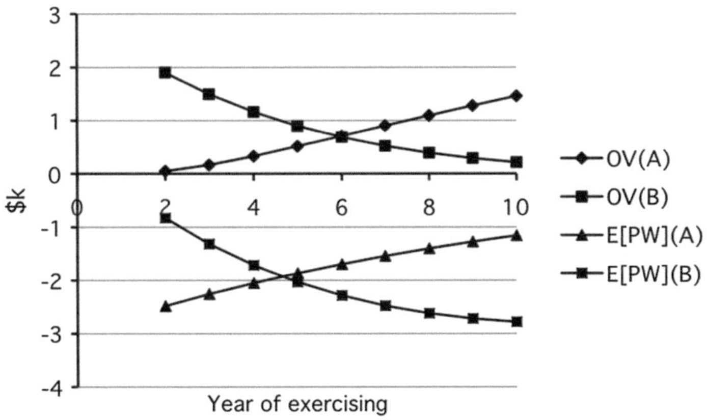

Fig. 3.9 Option value and expected present worth versus time; choice between two options (A and B)

3.7 Switch in Practices, Change in Form

Organizations may decide to change or switch the way they do things, for example contractually, or through changed method or changed resource usage, based on an option to switch or change. There may be a cost associated with this switching. The relevant cash flows to use in the option calculation are the difference between what cash flows would have existed (assuming no switching) and what new cash flows result from the switch (usually both positive and negative).

Convertible contracts. (Note, this is a reference to contracts as legally enforceable agreements, as distinct from the option usage of the term 'contract' above.) Consider a switch between contract payment types (here, cost reimbursable to lump sum) within a project, at year 1 of the project. The cost of doing this is based on the lump sum estimate, while the gain (cost foregone) is based on the cost reimbursable estimate. For the work remaining after 1 year, the cost reimbursable estimate is: $E[Y_{11}] = \$680k$ and $Var[Y_{11}] = (\$56.7k)^2$, where Y_{11} refers to cost reimbursable values; while the lump sum estimate expected value and variance are, respectively, \$650.0k and $(\$32.5k)^2$, giving: $E[Y_{12}] = (-)\$650.0k$ and $Var[Y_{12}] = (\$32.5k)^2$, where Y_{12} refers to lump sum values (Fig. 3.10). Since values for Y_{11} and Y_{12} are based on similar estimating principles, it is reasonable to assume that Y_{11} and Y_{12} are close to being perfectly correlated.

Present worth is based on discounting the difference between Y_{11} and Y_{12}, giving $E[PW] = \$27.3k$ and $Var[PW] = (\$81.1k)^2$. Equation (1.1) gives $OV = \$45.1k$ (approximately, 7% of the contract sum). This is the value of having convertibility within the contract. With a prescribed definite switch in payment types within the contract (compared with the option to switch as just calculated), this leads to a lower value (namely $E[PW] = \$27.3k$ or 4% of the contract sum).

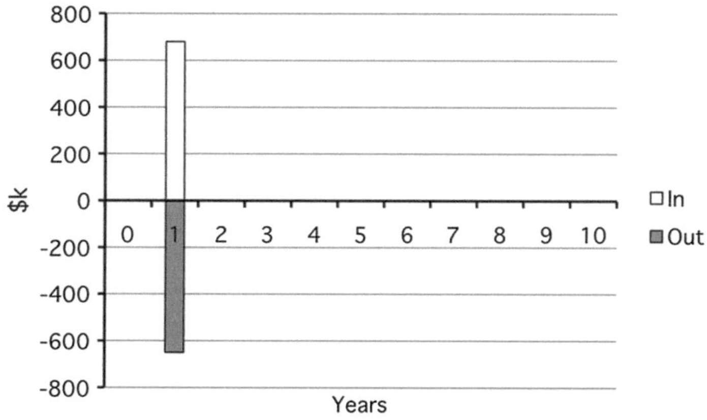

Fig. 3.10 Example—convertible contract; expected values of cash flows

Adaptable infrastructure. Adaptability related to infrastructure may be thought of in terms of: (i) Designed-in adaptability versus non designed-in adaptability; and (ii) Preset adaptation versus optional adaptation. With designed-in adaptability, the asset is deliberately designed, from the beginning, with the view that adaptation could (but not necessarily) take place in the future. With preset adaptation, explicit dates for adaptation may be set, and/or explicit adaptation may be prescribed.

Consider, for example, a seawall in the light of possible future climate shift and progressive sea level rising, both of which have uncertainties arising from multiple sources—environmental, human impact, social, political and so on. Climate shift projections, and resulting sea level rises, are informed approximations because of these uncertainties. Adaptation implies raising the seawall over time, in order to afford continual protection of adjacent inhabited regions.

For possible adaptation at a time T, based on sea level forecasts (Fig. 3.11):

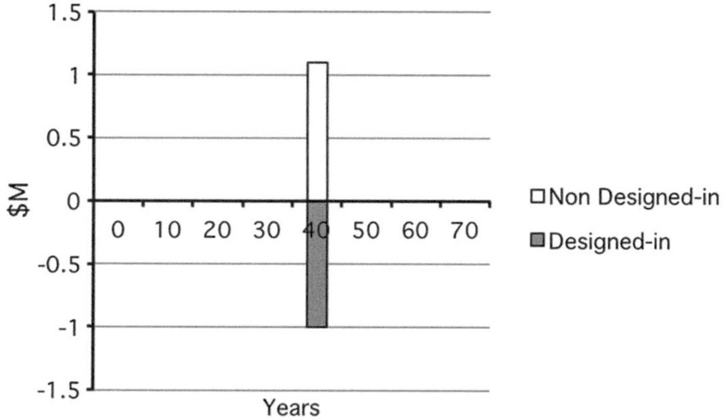

Fig. 3.11 Example—adaptable infrastructure; expected values of cash flows

(i) Designed-in adaptability versus non designed-in adaptability.
Let the cost of adaptation for the non designed-in version at T be Y_{T1}, and let the cost of adaptation for the designed-in version be Y_{T2}. Both Y_{T1} and Y_{T2} are random variables. This includes the situation where Y_{T1} corresponds to a completely new seawall being built.
(ii) Preset adaptation versus optional adaptation.
Let the cost of the adaptation (preset) at T be Y_{T1}. And let the cost of the adaptation (optional) be Y_{T2}. Both are random variables, but with Y_{T1} having a low variance.

For both (i) and (ii), the cost difference is examined; let $X_T = Y_{T1} - Y_{T2}$ (Fig. 3.11). The expected value and variance of X_T and PW follow. OV is obtained from Eq. (1.1). For viability in Case (i), OV is compared with any additional initial cost involved with designing-in adaptability features.

For Case (i), Y_{T1} and Y_{T2} will have similar variances and will be close to being perfectly correlated. This will lead to Var[X_T] being small, Var[PW] being small, and the majority of the present worth distribution occurring over positive PW values. Whether the designed-in version is viable will thus rest on E[PW] having to be greater than the additional initial cost involved in building-in adaptability. Even though future uncertainty is involved, it should be sufficient to only consider deterministic costs in a comparison of designed-in adaptability versus non designed-in adaptability.

For Case (ii), Y_{T1} and Y_{T2} could be anticipated to have similar expected values, that is Φ will be approximately 0.5, and hence optional adaptation will always be better than preset adaptation.

3.8 Delay, Deferment

The exercising of an option might be delayed or deferred until circumstances differ. The delay might be in anticipation of higher future market prices or advances in technology, or while waiting for regulatory approval. Delaying the exercising of an option, because of the resultant effect on cash flows, may lower or increase the option value, dependent on the circumstances.

The analysis of options involving delays is no different to the earlier plain or single options. All that needs to be done is to acknowledge the impact of the delay on the cash flows used in the analysis. This is so whether the project life, and hence the extent over time of the ensuing cash flows, is finite or very long. Delaying the exercising of an option may result in cash flows being lost, delayed or advanced, modified (increased or decreased), remaining as they were or a combination of these, dependent on the circumstances. But still the form of the analysis remains the same. Any additional direct cost (or gain) associated with a delay becomes yet another cash flow.

In establishing when is the best time to exercise an option, different available times can be considered in the analysis and option values compared via enumeration.

Examples A delay or deferment might result in cash flows being lost, delayed or advanced, modified, remaining as they were or a combination of these.

Consider an illustration. An expansion at year 2 costs: $E[Y_{22}] = -\$100k$ and $Var[Y_{22}] = (\$5k)^2$. This generates income each year following the expansion of: $E[Y_{i1}] = \$25k$ and $Var[Y_{i1}] = (\$6k)^2$, $i = 3, 4, \ldots, 7$ (Fig. 3.12). All cash flows are assumed strongly correlated. Discounting gives E[PW] and Var[PW], and Eq. (1.1) gives OV.

Three scenarios, with examples, can be considered:

(i) Cash flows lost. In delaying the expansion, the cash flow Y_{i1} in any year is consequently lost. For example, if the expansion is delayed 1 year, then Y_{31} is lost (Fig. 3.13).

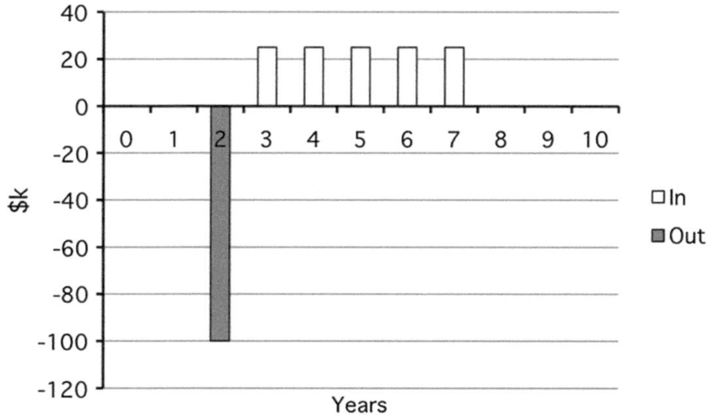

Fig. 3.12 Example—original expand option prior to delay; expected values of cash flows

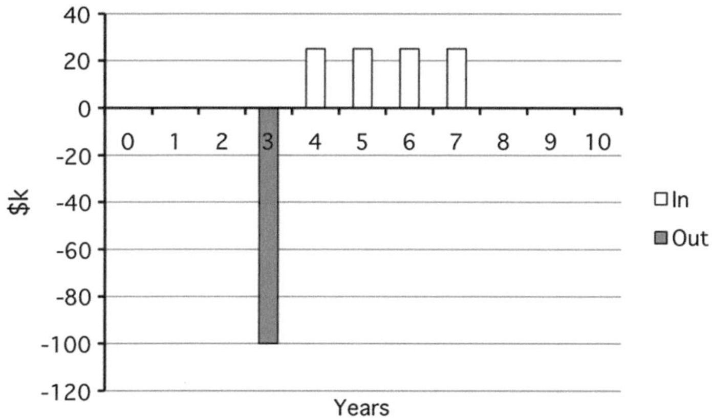

Fig. 3.13 Example—expand option, delay 1 year, cash flows lost; expected values of cash flows

(ii) Cash flows delayed or advanced. In delaying the expansion, the cash flows Y_{i1} in the following years are delayed. For example, if the expansion is delayed 1 year, then Y_{i1} now applies over i = 4, 5, …, 8 (Fig. 3.14).

(iii) Cash flows modified. In delaying the expansion, the expected values and variances of all the cash flows in the following years increase: for 1 year delayed, 5%; for 2 years delayed, 10%; and so on (Fig. 3.15).

Figure 3.16 shows the example option values versus years delayed for the three above-given scenarios.

For delay analysis, Fig. 3.16 applies for a particular set of cash flows and cash flow assumptions, and is not intended as general behaviour associated with delay.

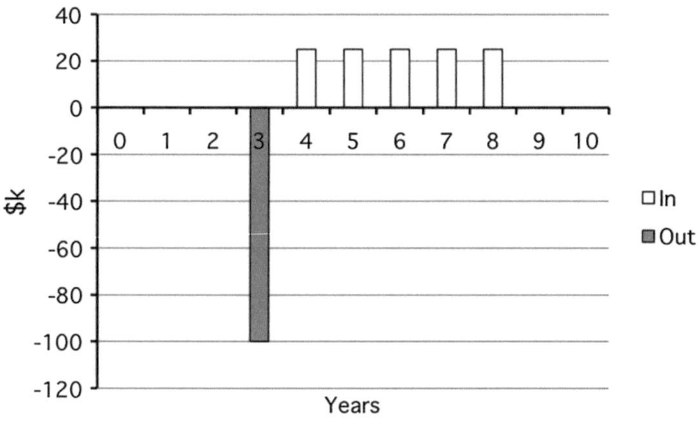

Fig. 3.14 Example—expand option, delay 1 year, cash flows delayed; expected values of cash flows

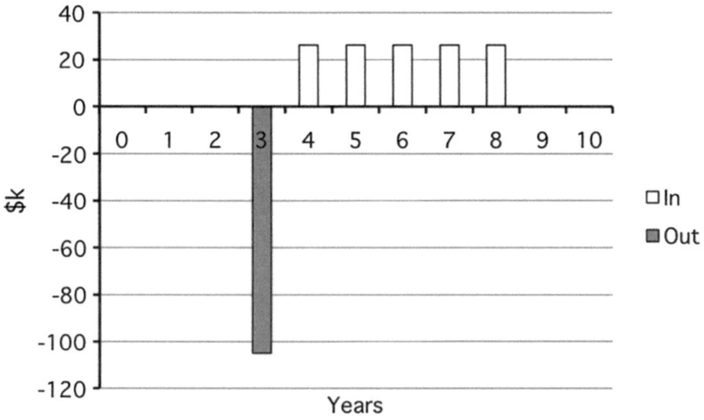

Fig. 3.15 Example—expand option, delay 1 year, cash flows increased; expected values of cash flows

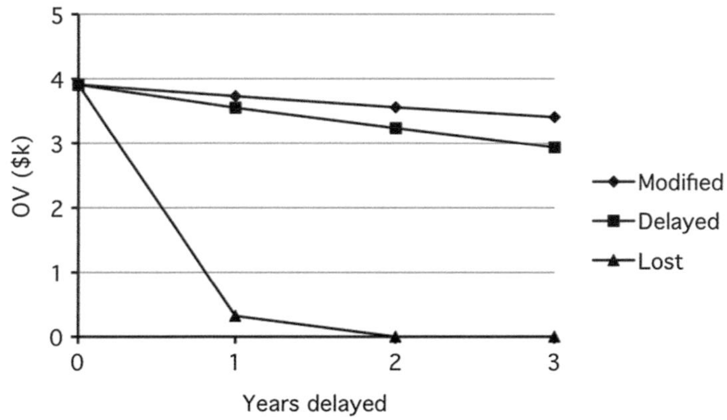

Fig. 3.16 Example delay analyses; option value versus days delayed

Generally, with traditional financial options, delay in exercising is rewarded, while with real options, delay may be penalized or rewarded.

3.9 Sequential Options

Sequential options are a chain of options. For example, in a project comprised of stages, there may exist an option at each stage, and this determines what subsequently happens with the project. The options in a chain might be any of the single options, in different sequences, and hence there is no one standard sequential options case. In the sequence of options, naturally a preceding (independent) option is exercised before a succeeding (dependent) option.

Where the exercising of each option generates its own cash flows, the value of the following option is included in the valuation of the option at hand. This is in line with Bellman's principle of optimality. Such sequential options are analysed backward in time, but each option analysis is no different to the earlier single options.

Examples Two typical sequential option cases are presented here, related to whether: (i) exercising the component options generates their own cash flows, or (ii) all component options need to be exercised in order to generate a cash flow. The example treatments below apply generally to any number of multiple component options.

(i) Each exercise generates cash flows

Consider sequential options occurring at $t = t1, t2, \ldots$ where $t1 < t2 < \ldots$ The following (dependent) option can only exist if the preceding (independent) option has been previously exercised. Exercising the first option at t1 generates positive and negative cash flows reducible or equivalent to $Y_{t1,1}$ and $Y_{t1,2}$ respectively, exercising the second option at t2 generates positive and negative cash flows reducible or equivalent to $Y_{t2,1}$ and $Y_{t2,2}$ respectively, and so on up to the last option at time tq.

Bellman's principle of optimality stated in the 1950s is relevant here [2]; see [3], p. 139 and [4]: *An optimal policy has the property that whatever the initial state and initial decision are, the remaining decisions must constitute an optimal policy with regard to the state resulting from the first decision.*

Valuing the options in reverse order, the option value for the last option follows using $Y_{tq,1}$ and $Y_{tq,2}$ discounted to time tq–1, and is denoted OV_{tq}. The second last option is then valued based on cash flows $Y_{tq-1,1}$, $Y_{tq-1,2}$ and OV_{tq}, all discounted to time tq–2. This gives OV_{tq-1}. The procedure for the second last, third last etc. options is repeated in a backwards time sense, ending with the first option calculation, giving OV_{t1}, which is the option value for the sequential option.

R&D example. Consider a three-sequential option, manufacturing R&D example. The data follow. Product introduction stage: $E[Y_{t1,1}] = -250M$, $Var[Y_{t1,1}] = (\$62M)^2$, $E[Y_{t1,2}] = \$182M$, $Var[Y_{t1,2}] = (\$45M)^2$; First expansion stage: $E[Y_{t2,1}] = -\$250M$, $Var[Y_{t2,1}] = (\$62M)^2$, $E[Y_{t2,2}] = \$245M$, $Var[Y_{t2,2}] = (\$61M)^2$; Second expansion stage: $E[Y_{t3,1}] = -275M$, $Var[Y_{t3,1}] = (\$68M)^2$, $E[Y_{t3,2}] = \$322M$, $Var[Y_{t3,2}] = (\$80M)^2$ (Fig. 3.17). The cash flows at t1 = 1, t2 = 2, and t3 = 4 years are assumed independent.

Each of these is an expand option, and calculated as discussed above. OV_{t3} is calculated first and equals \$56.4M. This is included in the calculation of OV_{t2}, which equals \$59.0M. This is included in the calculation of OV_{t1}, which equals \$31.6M, the option value for the total sequential option.

CDM projects. Clean Development Mechanism (CDM) projects can choose to have a 7-year crediting period that can be renewed twice, giving a total of 21 years [1, 8]. The choice of an initial 7-year crediting period, and the subsequent renewal or non-renewal for a following 7 or 14 years, can be seen as a sequential option, that is an option to discontinue or continue as a CDM project at the end of year 7, and later possibly at the end of year 14. Option calculations only consider cash flows that are

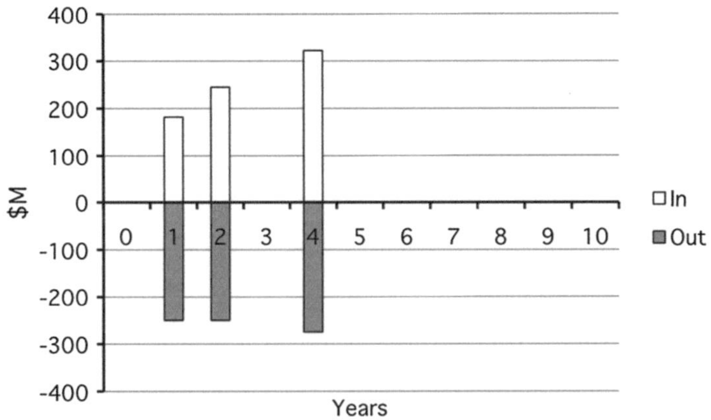

Fig. 3.17 Example—sequential options, each component with cash flows; expected values of cash flows

CER- or carbon-based (inflows and outflows) and not all project cash flows. (CERs are carbon emission reduction units, or carbon credits) The analysis of the option occurring at the end of year 7 (assuming renewal only for one more 7-year period) is calculated in terms of the anticipated carbon-based costs (CDM operational costs) and anticipated carbon-based revenues (based on forecast prices and project output of CERs) for years 8 to 14. All these costs and revenue contain uncertainty. This option value then needs to be compared with its 'premium', namely the (additional) cost of preparing updated documentation at the end of year 7. The option value would need to exceed this additional administration cost in order to proceed as a CDM project. That is, there could be value in having flexibility to either discontinue or continue a project at the end of year 7, dependent on anticipated costs and anticipated revenue at year 8 and over the following years up to year 14.

Similarly, if the choice at the end of year 7 is that of continuing CDM registration, there exists an option at the end of year 14.

The total option value of having options at the end of both years 14 and 7 is evaluated by working backwards in time. The option value at the end of year 14 is calculated based on the cash flows in years 15 to 21. This option value, together the additional administration cost at year 15 and the cash flows in years 8 to 14, give the option value at the end of year 7. This option value, together the additional administration cost at year 8 and the cash flows in years 1 to 7, give the option value at year 0. This is the total option value and is compared with the up-front CDM cost, in order to establish CDM viability.

(ii) All options needing exercising to generate cash flows

Consider sequential options occurring at $t = t1, t2, \ldots$ where $t1 < t2 < \ldots$ A following (dependent) option can only exist if the preceding (independent) option has been previously exercised. Exercising all options except the last generates no direct cash inflows, but may have cash outflows reducible or equivalent to $Y_{t1,2}$, $Y_{t2,2}$, ..., while exercising the last option at tq has a cash outflow reducible or equivalent to $Y_{tq,2}$, and cash inflow reducible or equivalent to $Y_{tq,1}$. The option value might be thought of as evolving in time, however, intermediate option values are not realizable, since it is only on exercising the last option that a return is gained.

Consider a manufacturing plant example. Three project stages are involved: a land acquisition and permitting stage, a design stage and a construction stage, starting at $t1 = 1$, $t2 = 3$, and $t3 = 5$ years respectively. Cash outflows associated with the options at these points in time are: $E[Y_{t1,2}] = -\$30M$, $Var[Y_{t1,2}] = (\$3M)^2$; $E[Y_{t2,2}] = -\$90M$, $Var[Y_{t2,2}] = (\$11M)^2$; and $E[Y_{t3,2}] = -\$210M$, $Var[Y_{t3,2}] = (\$26M)^2$, respectively. The cash inflow occurs at the end of the construction stage (7 years), giving a discounted cash inflow at t3 of: $E[Y_{t3,1}] = \$402M$, $Var[Y_{t3,1}] = (\$100M)^2$ (Fig. 3.18). All cash flows are assumed well correlated. The option value for the total investment is calculated by discounting all cash flows to the present, with Eq. (1.1) giving OV = \$29M.

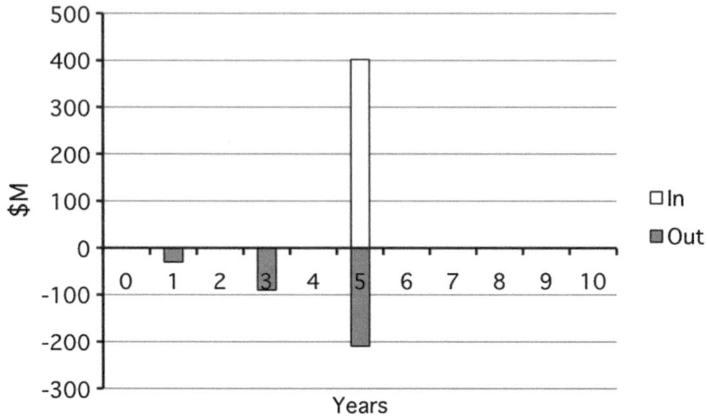

Fig. 3.18 Example—sequential options, all components needing exercising; expected values of cash flows

3.10 Parallel Options

Parallel options are characterized by a dependent option and an independent option, as in sequential options, but now applying over the same time frame, rather than in sequential time frames. The component options may be any of the single options. The independent option is exercised first. The analysis mirrors that of sequential options.

Parallel options have an independent option and a dependent option as for sequential options, but these component options exist over the same time frame. In principle, the valuation for parallel options is no different to that for sequential options. The same two cases examined in sequential options above, could apply to parallel options. For the two-component parallel option case, in line with the terminology used for sequential options, let the time frame for exercising be t1. The independent option is evaluated over t1, while the dependent option is evaluated over a time frame of $t2 = t1 + \Delta t$, where Δt is a small increment in time (Fig. 3.19). That is, by imagining that the independent option is exercised at a very small time interval different from the dependent option, then the same thinking as sequential options applies, and the option analysis, again, is no different to the earlier expand and contract options. In effect, for calculation purposes, $t2 = t1$.

With compound options, because the difference between the estimates of the two forms of analysis can be both positive and negative, depending on the values taken by the variables, it is possible that the component option value estimates may balance each other out, rather than enlarge the difference between the estimates of the two forms of analysis. However, no general conclusion on this is possible.

With compound options, the existing literature works on traditional financial option-style examples using volatility and present day asset values, and hence there is little that the book's approach can directly compare to. The values of individual options, within the compound options, are consistent with Black-Scholes, and the

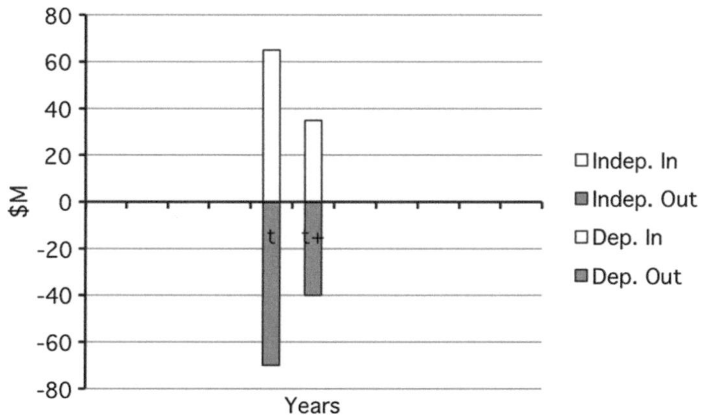

Fig. 3.19 Example—parallel options; expected values of cash flows

compound option approach is consistent with the existing literature, and so it is anticipated that the option values calculated will be satisfactory. Where comparisons are able to be made between the two forms of analysis, the trends in compound options, with respect to Φ, time of exercising, interest rate and uncertainty are similar to those trends identified above for plain options.

3.11 Rainbow Option

A rainbow option refers to there being more than one type of uncertainty present. For example in mining, the uncertainty in the cash flows can come from the price of the ore, ore yield and the mine output. While this prevents or complicates the calculation if a traditional financial options method is used (for example, a binomial tree goes to a quadrinomial tree where there are two sources of uncertainty), the calculations remain essentially the same if the analysis is using cash flows. The calculations are increased only slightly. All uncertainty sources are assembled, acknowledging any correlation present, into the estimates used for cash flows. Having incorporated all the uncertainty sources into the cash flows, the analysis proceeds as for any of the other options.

For example, consider a cash flow Y made up of price, Z_a, and output, Z_b, both random variables, in the form,

$$Y = Z_a Z_b$$

Then,

$$E[Y] = Cov[Z_a, Z_b] + E[Z_a]E[Z_b]$$

and for Z_a and Z_b independent,

$$\mathrm{Var}[Y] = E^2[Z_a]\mathrm{Var}[Z_b] + E^2[Z_b]\mathrm{Var}[Z_a] + \mathrm{Var}[Z_a]\mathrm{Var}[Z_b]$$

Other expressions are available for different assumptions on the correlation of the variables containing uncertainty, as well as for more than two sources of uncertainty [5].

3.12 Closure

It is believed that the book's cash flow approach will make real options analysis more commonplace and more accessible to a greater numbers of people. In this form, it can be given as an add-on to conventional discounted cash flow analysis taught to undergraduates; no new thinking is required.

The cash flow approach to real option estimates is very straightforward, requiring minimal mathematical background. The approach does not distinguish between the type of option, for example whether it is an expand-style option, contract-style option, involves delays or contains multiple sources of uncertainty (correlated or not); the approach is the same for all options. It uses variance instead of volatility, a term which does not have meaning in most real options, and avoids trying to establish an analogous financial option in order to force fit established financial options methods. Any set of cash inflows and cash outflows, whether correlated or not, can be accommodated through the one approach. The present worth can take both positive and negative values; there are no inherent lognormal assumptions, rather the distribution for PW can be selected at the discretion of the person doing the calculations. And if a second order moment approach is adopted, there are no distribution assumptions necessary on the cash flows. Interest rates and rate mixes can also be chosen at the discretion of the person doing the calculations.

Heretofore, real options have been valued by adopting analogies with traditional financial options, and the number of technical publications using such methods is very large. Something new and different, in general across all human endeavours, may not be well accepted by people. Coupled with the book's cash flow approach being very simple, it is envisaged that there will be initial public resistance to adopting or public unwillingness to move to a cash flow approach.

For plain options, the relationship between the option value and the analysis input variables of Φ, time to exercising, uncertainty, and interest rate is similar between the approach and that of Black-Scholes. The book's approach gives essentially the same option values as Black-Scholes, as also observed in Carmichael et al. [7]. The larger differences with Black-Scholes occur with higher values of: Φ, time to exercising, interest rate and uncertainty. However, anticipated real option applications will have lower values of Φ, time to exercising, interest rate and uncertainty than those at which the book's approach starts to diverge from Black-Scholes. For compound options, where comparisons are able to be made, similar trends are observed.

Although the chapter compared option values from the book's approach with Black-Scholes, for a restricted number of cases, it is remarked that there is no 'correct' answer to compare the real option value with. It is only possible to qualitatively compare with methods such as Black-Scholes, because different situations are being looked at in real options compared to traditional financial options. However, the restricted comparisons show that the book's approach produces reasonable values.

The book's cash flow view of real options provides an intuitive user-friendly approach, requiring minimal mathematical background, devised to specifically value real options and aligned with usual investment calculations, without the need for financial options analogies, constraints, terminology and variables such as volatility. Accordingly, the cash flow view should be acceptable to most.

3.13 Extensions

The chapter provides the skeleton for a cash flow approach to valuing real options. Future developments could examine possible refinements, for example in the choice of the present worth distribution and assumed interest rates, and in examining special-case compound options.

References

1. Baker and McKenzie (2015), *CDM Rulebook*, http://www.cdmrulebook.org (Viewed 31 January 2015)
2. Bellman RS (1957) Dynamic programming. Princeton University Press, Princeton
3. Carmichael DG (1981) Structural modelling and optimization, Ellis Horwood Ltd. (John Wiley and Sons), Chichester
4. Carmichael DG (2013) Problem solving for engineers. CRC Press, Taylor and Francis, London
5. Carmichael DG (2014) Infrastructure investment: an engineering perspective. CRC Press, Taylor and Francis, London
6. Carmichael DG (2016) A cash flow view of real options. Eng Econ 61(4):265–288
7. Carmichael DG, Hersh AM, Parasu P (2011) Real options estimate using probabilistic present worth analysis. Eng Econ 56(4):295–320
8. Carmichael DG, Lea KA, Balatbat MCA (2015) The financial additionality and viability of CDM projects allowing for uncertainty. Environ Dev Sustain 18(1):129–141

Chapter 4
Adaptable Infrastructure—Civil

4.1 Introduction

Infrastructure over its lifetime needs to consider change, often because of external forces, in order to avoid obsolescence in any of its various forms—physical, economic, functional, technological, social, environmental and legal. Demographic, technological and environmental shifts would appear inevitable, but to an uncertain extent and with uncertain timing. This leads to considering flexibility and the adaptation of infrastructure in line with shifts, and the comparison of two alternative approaches:

- Where adaptability features have been designed and built in ab initio, with the view that adaptation may (but not necessarily) take place in the future depending on future circumstances. The infrastructure has built-in flexibility. (Termed here: A form.) That is, there is an allowed-for ability to adapt in the future, should it be required, but this adaptation need not be done. There is an ability to do something in the future but not an obligation to do this.
- Where the infrastructure has been designed and built without adaptability features in mind, but where future adaptation may still be fortuitously possible, albeit with greater effort. (Termed here: NA form.) This is business-as-usual (BAU).

Heretofore, there has not been presented a general framework that can be used to establish whether building in flexibility is better or worse than not doing so. This is the subject of this chapter. 'Better or worse' here is interpreted in the sense of sustainability criteria, namely social, environmental and financial. The framework presented here allows for uncertainty in the extent and timing of any, often externally forced, change.

Adaptation is viewed in terms of an option, namely a right but not an obligation to adapt. The option to adapt may never be exercised. Having this option leads to greater value over the case of certainly adapting. Social and environmental issues are

D. G. Carmichael, *Future-proofing—Valuing Adaptability, Flexibility, Convertibility and Options*, Management in the Built Environment,
https://doi.org/10.1007/978-981-15-0723-6_4

converted to monetary values such that the analysis with respect to social, environmental and financial criteria can be done together, though converting intangibles to monetary values is not without its critics.

The chapter firstly reviews the topic. This leads to an outline of the chapter's general analysis. The chapter's approach is free of any particularities of infrastructure or change type, only conversing in costs, benefits and disbenefits. Later examples on roads demonstrate the applicability of this analysis.

4.2 Examples of Possible Built-in Adaptability/Flexibility

There are many existing examples of flexibility and forethought within infrastructure. The literature on flexibility in infrastructure is growing. Adaptability/flexibility thinking is not new. Some examples are given here.

Roads

For future increased demand:

$t = 0$: Allow for a wider road reserve; or perform the earthworks for extra lanes in each direction; or construct the sub-base for extra lanes, but do not seal; or construct extra bridge abutments where the road crosses a river, another road or rail line; or construct a larger or extra tunnel but do not fit out.

$t = T$: Increased lanes.

Seawalls

For future shifts in seas:

$t = 0$: Design/build to facilitate future extension with increase in sea level/waves.

$t = T$: Increased seawall height.

This chapter shows how such adaptability/flexibility can be valued.

4.3 Flexibility Comparative Analysis

The comparative analysis given here can be applied to any deliberate built-in flexibility situation. Two forms of adaptation (A—deliberate built-in flexibility that facilitates adaptation, and NA—no built-in flexibility, the more usual practice, or business-as-usual—BAU—practice) are considered and compared in financial, environmental and social terms. The analysis considers infrastructure (built at time $t = 0$), with the potential to be adapted at some time $t = T$ in the future. There is a trade-off between financial, environmental and social impacts now and in the future. The adaptation, however, need never occur, depending on the infrastructure owner's situation in the future. That is, the adaptation is an option—a right but not an obligation. The time, T, can be varied in the calculations in order to show the relationship between time of any adaptation and flexibility value.

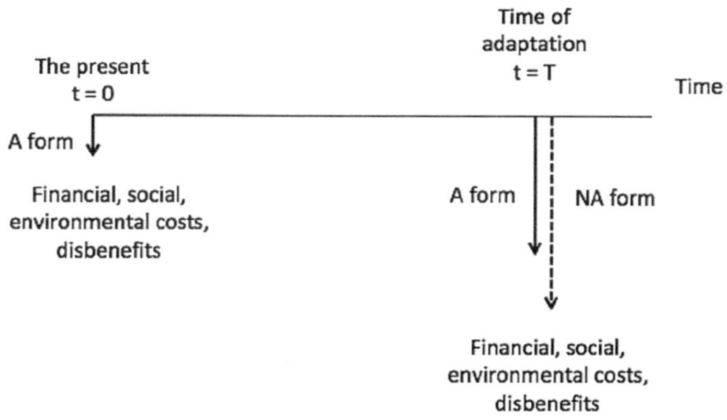

Fig. 4.1 Cash flows for comparative analysis

To establish the option value (but not the complete viability of flexibility), only costs and disbenefits (expressed in dollars) at T are taken into account. Benefits of the NA and A versions are assumed to be the same post-adaptation, and do not enter the analysis (but should they be different, then they too can be incorporated in the same way, but with an opposite sign). Any costs and disbenefits post-adaptation, common to NA and A, also do not need to enter the analysis. Figure 4.1 shows the overall cash flows of relevance.

Impacts between $t = 0$ and $t = T$ are assumed to be the same for both A and NA forms. Should this assumption not be so in particular cases, the NA-A differences can be included in the analysis. Desirably, well-designed A form infrastructure leads to lower environmental and social impacts, contributed to by lesser construction site time and effort in the adaptation. Different environmental and social issues need to be considered for each specific application.

To ascertain the value of flexibility (A form) compared with more conventional (BAU) practice (NA form), the difference between the NA and A forms is examined. The net costs and disbenefits at T, X_T, is defined as in Appendix 2.11.1, where $i = T$. In a comparative analysis, because the difference between the NA form and the A form is being looked at, costs and disbenefits, Y_{Tk}, $k = 1, 2, 3, \ldots, \mu$, of the NA form are considered positive, while costs and disbenefits, Y_{Tk}, $k = \mu + 1, \mu + 2, \ldots, m$, of the A form are considered negative, and X_T is the difference between the NA and A forms. Strict accounting conventions are not followed. Rather, everything is looked at from the viewpoint of the person doing the calculations, here taken as the investor or option holder, such that something favourable to the investor (avoiding the costs and disbenefits of the NA form at time T) is taken as positive and something unfavourable to the investor (having the costs and disbenefits of the A form at time T) is taken as negative.

Because estimates for the NA and A forms are based on similar assumptions, it could be anticipated that there would be very strong correlation between the estimates used for the NA and A forms.

Present worth then follows from Appendices 2.11.2 or 2.11.3. And the flexibility value is given by Eq. (1.1).

This flexibility value is then compared with the cost of building in flexibility at time 0. Viability is established for flexibility when the flexibility value exceeds this initial cost.

4.4 Summary of Method

The following summary method can be used by those wishing to establish the viability of building in flexibility into infrastructure:

1. Costs and disbenefits at $t = 0$ of the NA and A forms are estimated. The difference is the additional premium involved in building in flexibility.
2. Expected values and variances of the costs and disbenefits at $t = T$ (the time considered for adaptation) of the NA and A forms are estimated.
3. These expected values and variances are discounted to $t = 0$, to give E[PW] and Var[PW] from Appendix 2.11.2 or 2.11.3.
4. A probability distribution, selected at the discretion of the person doing the calculations, is fitted to PW.
5. The value of flexibility, ΦM, is calculated (Eq. 1.1).
6. This flexibility value is compared with the premium calculated in item 1.

4.5 Case Example

The valuation of flexibility in extending the number of lanes of an arterial road is demonstrated here [4].

Relevant costs and disbenefits that need separate valuation at $t = 0$ and T, for both the NA and A forms, are related to: (a) land resumption and construction costs; (b) emissions, pollutants; (c) waste; (d) employment; (e) safety, health; (f) noise; (g) inconvenience (delays, disruptions, relocations); (h) other environmental impacts; and (i) other social impacts. Reference below to (a), (b) and so on, is to these costs and disbenefits. Items (b) to (i) only become disbenefits when they appear as negative consequences, for example reduced employment or increased possibility of accidents.

Assume for definiteness, based on estimated existing (and near future) vehicle demand, that two lanes in each direction is sufficient for $t = 0$. Also assume, based on a later estimate (made at $t = T$) of demand at T, that the road now requires three lanes in each direction. (However, note that the following applies to any assumption on the number of lanes, at $t = 0$ or T.) Different T values can be used in the calculations in order to establish how the value of flexibility varies over time.

t = 0

NA form: The road is constructed with two lanes in each direction, and without any forethought of changing the road in the future. This might be called the business-as-usual (BAU) case.

A form: The road is constructed with the view that increased future demand, beyond that used in the NA form, may occur. This consideration may lead to thought about a range of possible in-built adaptation features, such as (singly or in combinations, relative to the BAU case): allowing for a wider road reserve; locating noise barriers and landscaping further away from the road; constructing the supports of overpasses (going over the road) so as to facilitate the placement of extra lanes for the road; performing the earthworks for more than two lanes in each direction; constructing the sub-base for more than two lanes in each direction (but only sealing two lanes in each direction); constructing extra bridge abutments where the road crosses a river, another road or rail line; or constructing a larger or extra tunnel.

All the above make use at t = 0 of already-mobilized equipment and people, rather than having to remobilize at some later date T. Clearly, the sum of (a) to (i) for the A form is greater than the sum of (a) to (i) for the NA form at t = 0. The difference, or premium paid, between the NA and A forms is noted.

t = T

NA form: The cost and disbenefits involved are effectively those related to constructing a new one-lane road in each direction.

A form: An extra lane in each direction is still required, but the preparatory work is now considerably reduced, and the road reserve already exists to accommodate the extra lane.

Both NA and A forms are hampered by maintaining existing traffic flow and providing a safe work and driving environment. These add extra to the construction cost, but are less for the A case.

Clearly, the expected sum of (a) to (i) for the A form is less than the expected sum of (a) to (i) for the NA form at t = T. The difference between the NA and A forms (incorporating both expected values and variances), when discounted to t = 0, is used to value the flexibility, equivalently OV, according to Eq. (1.1).

Overall viability of building in flexibility will depend on a comparison of the premium paid with OV. Where OV exceeds the premium, the introduction of flexibility is worthwhile.

4.6 Closure

The chapter gave a general method for valuing built-in flexibility in infrastructure. It did this by comparing infrastructure without and with built-in flexibility. Applications were outlined where flexibility could be entertained. The method given in this chapter is independent of any particular infrastructure, and particular change, or adaptation envisaged. The author's experience is that, in some cases, having flexibility will be viable, while in other cases it will not; different assumptions and different geographical locations will lead to different viability conclusions. Infrastructure practices are location-dependent, and there are too many variables to enable generalized conclusions to be made across all infrastructure, all adaptations, and all locations. Whether flexibility is viable in any particular case will require a case-by-case analysis.

Initial calculations suggest that building in flexibility is viable where adaptation is considered in the short term (for example, a road upgrade within the first 25 years of the life of a road), where interest rates are low, and where uncertainty about the future is large [1]. The social and environmental considerations increase viability over purely financial considerations, and hence having flexibility would be more viable and appeal more to public sector infrastructure owners rather than the private sector. At present, many in the private sector place little weight on social and environmental issues.

Flexibility has some value, however it is unclear in any infrastructure sale, what value the market might place on any additional value offered by flexibility, and in particular it would be difficult for the seller to establish what value the buyer places on this flexibility.

The architecture literature suggests that there is value in flexibility, but has stopped short in quantifying this value, while the civil infrastructure literature has attempted the use of traditional financial options analogies to establish value. However such analogies have been called into question, in particular with establishing equivalent volatilities, which do not exist in infrastructure, deterministic exercising and other modelling assumptions [2]. This chapter provides a rational analysis that gives the value of any flexible infrastructure. The approach will be useful to infrastructure owners contemplating prolonging the useful life of infrastructure through alteration over time.

A further argument in favour of built-in flexibility may be an aesthetic one. Adaptation to infrastructure with built-in flexibility could be anticipated to blend in more with the existing infrastructure, when compared to the no built-in flexibility case. Flexibility considerations may also feed into risk management studies [3].

Some general conclusions

- Building-in adaptability/flexibility involves some extra initial cost, which can be minimised through good engineering. But this may deter those who want cheapest initial cost.
- Building in adaptability is most attractive where adaptation is anticipated in the shorter term (for example, a fast growing population). The attractiveness of longer-term adaptation (for example, climate shift) will depend on rates and initial costs; using environmental/social discounting will make it more viable.
- Built-in adaptability will be more viable for the public sector than the private sector because of the lower rates and intangibles used in feasibility studies.
- Social and environmental issues add to the value of building in adaptability, over pure financial arguments.

- The 'ingenuity' of engineers can overcome hurdles where there is no forethought to adaptability.
- The main arguments in favour of building in adaptability may be aesthetics and function.

4.7 Extensions

The engineering aspects of flexibility need to be explored, such that minimal expense occurs at $t = 0$. This will involve creative thinking. This could make built-in flexibility features more commonplace or even the future norm in infrastructure. Commonly, owners have only restricted access to finance at $t = 0$, or wish cheapest initial cost.

The method given in this chapter has broader applicability than infrastructure, and extends to flexibility analysis generally, for example to urban 'planning'.

References

1. Carmichael DG (2015) Incorporating resilience through adaptability and flexibility. Civ Eng Environ Syst 32(1–2):31–43
2. Carmichael DG, Hersh AM, Parasu P (2011) Real options estimate using probabilistic present worth analysis. Eng Econ 56(4):295–320
3. Carmichael DG (2016) Risk–a commentary. Civ Eng Environ Syst 33(3):177–198
4. Carmichael DG (2018) Estimating the value of built-in flexibility in infrastructure. In: Wang C, Harper C, Lee Y, Harris R, Berryman C (eds) ASCE construction research congress 2018: sustainable design and construction and education, ASCE, New York, pp 613–622, ISBN (PDF) 9780784481301, ASCE Construction Research Congress, New Orleans, Louisiana, 2–5 April 2018

Chapter 5
Adaptable Buildings and Houses

5.1 Introduction

Flexibility is seen as an important issue within modern commercial and residential buildings, as the needs and wants, real and perceived, of the inhabitants transition over time. With houses, the needs and wants of inhabitants shift in line with their lifestyle and employment/income/family/schooling situation and social factors. With buildings generally, change may come about through shifts in technology, the market and the economy. However, buildings are typically built for a single purpose, which suits the owner at the time. With houses, as families grow/shrink, house owners look for more/less space. This might be accomplished by relocating to another neighbourhood or, if this is not desired, converting the current house, including the undertaking of alterations and additions, with the extreme being the sustainability-unattractive complete demolition and replacement. Ever-present shifts and transitions imply, subject to demonstrated viability, that there may be a need in the market for buildings with deliberate in-built flexibility.

The literature on flexibility in buildings is growing, but argues qualitatively, and largely from an architectural perspective in favour of flexibility. There, the literature focuses on such matters as changing interiors of buildings, changing the building space volume, and changing building use. The architecture literature acknowledges a need for flexibility in order to delay obsolescence. However, the literature stops short in quantitatively valuing deliberate built-in flexibility. This chapter provides an approach for valuing deliberate built-in flexibility. Financially, the value is established through an options analysis, while social and environmental matters are evaluated along life cycle assessment lines. Financial, social and environmental considerations are looked at singly and in combination. Some numerical studies are given. Building conversion is done in response to shifts in usage and requirements over time. It is acknowledged that many types of change or building alteration are possible, and that building practices are location-dependent, leading to too many variables to enable any generalisation. However, the approach given in this chapter is general;

D. G. Carmichael, *Future-proofing—Valuing Adaptability, Flexibility, Convertibility and Options*, Management in the Built Environment,
https://doi.org/10.1007/978-981-15-0723-6_5

it carries over to all situations and locations, though the actual numbers used and building practices do not.

Two forms of alteration (denoted A and NA here) are considered and compared:

A form. Where flexibility features have been deliberately designed and built in ab initio, with the view that alteration may (but not necessarily) take place in the future depending on future circumstances.

NA form. Where a building has been designed and built without flexibility features in mind, but where future alteration may still be fortuitously possible, albeit with greater thought and effort.

The chapter will be of interest to people within the building and housing industries, as well as building and house owners. Through the approach given, it is possible to gauge the viability of including specific and deliberate built-in flexibility in any design and construction. The chapter applies to the adaptability and conversion of all building types, such as houses, offices, warehouses and other commercial and industrial buildings.

The chapter comments on the literature on usage and demand shifts. Numerical studies are presented and an argument on the viability (broadly from a sustainability viewpoint) of deliberate built-in flexibility given. The chapter does not look at the specifics of design or construction, but rather the valuation based on design concepts. Nor does it examine any constraints on building design or work that may be imposed by government regulations. This chapter gives a rational quantitative analysis for the financial value of flexibility, with social and environmental matters evaluated along life cycle assessment lines. The approach is free of any particularities of building or change type, but naturally the specific numbers differ across applications.

5.2 Examples of Possible Built-in Adaptability/Flexibility

5.2.1 General

There are many existing examples of flexibility and forethought. Adaptability/flexibility thinking is not new. This chapter shows how such adaptability/flexibility can be valued.

Adaptability may be designed and built into a building in many ways, and with creativity the extra cost and disbenefit impact can be minimised [1]. In many applications, the labour and equipment costs will be no different to the situation where adaptability has not been considered. Some examples and issues involved in adaptable design are discussed here.

5.2.2 *Houses*

For future extra space needs (growing family, …):

t = 0 Design/build one-storey house for possible future house 'extension'.
t = T Two-storey house.

For future less space needs (children leave home, …):

t = 0 Design/build house for possible future house 'division' to a duplex or semi-detached houses.
t = T Duplex style building.

One- to two-storey houses

Knowing that the ceiling of a one-storey or single-level house will become the floor of the storey above, timbers that can withstand floor loads can be used for the ceiling. These ceiling timbers could be covered with flooring sheets, and the roof mounted on this. With a roof designed to be reused (assuming the same footprint for both levels), the second level becomes one of raising the existing roof, and adding second level walls to support. Materials reuse is maximised, and waste minimised. There may be no need for the house owner to move out while new construction takes place. It is assumed that first level walls and footings are capable of supporting a second level. Room layout is chosen to facilitate the introduction of stairs and maintain good people movement around the house.

An alternative is that the first level be jacked up to become the second level, and a new first level built underneath.

Current practice in adding a second level is quite messy, usually involving the house owner moving out while deconstruction and construction takes place. New timbers are inserted on blocks to support the new second level, and this leads to a visually unattractive height between the ceiling below and the floor above. A new second level, including a new roof, is added and the old roof is discarded. Time on site at the time of adaptation is large because of the issues involved in changing something that was not designed to be changed. There is almost no materials reuse, but much waste. Stairs are inserted where it is possible to do so, but this may not suit the people movement in the house. The house always looks as though it has had a second level extension as an afterthought, and also is visually unattractive in many modified houses.

Current renovation and addition/extension costs per square metre are nearly twice that of new construction. With a house designed for adaptation, the cost per square metre could be anticipated to be closer to that of new construction. This represents a considerable saving, and with social and environmental (and generally, sustainability) issues additional, it would make design for adaptation a better choice than the sustainability-unattractive 'knock-down, rebuild' that is promoted by some builders.

Single dwelling to duplex

The principal features of a duplex or semi-detached house are that they are separated by soundproof and fireproof walls, and that they have separate entrances. To design for future division of a house thus requires minimal initial extra costs and disbenefits—that associated with the additional dividing wall material, and secondary entrances/exits. Room layout is chosen with two uses in mind—a single house owner, and two house owners. At time of adaptation, the separating firewall is made complete.

Without design for adaptation, the new construction (separating wall and entrances) can be messy, and the room layouts do not lend themselves to people movement.

5.2.3 Buildings

For future extra height, extra space:

$t = 0$	Allow extra in the footings and columns.
$t = T$	Extend height of building.

For future internal layout changes:

$t = 0$	Design without internal walls/columns.
$t = T1, T2, T3, \ldots$	Change the internal layout.

Open plan buildings

The intent of open plan buildings is to allow flexibility for internal future changes. Floor plans are designed to contain large open spaces. Within a commercial building, open plan allows reconfiguring work flows or changing internal room layouts in response to shifting technology and markets. Within offices, open plan may be praised for attempting to promote collaboration among workers, or criticised for denying workers privacy and quiet, and reducing productivity. Nevertheless, open plan allows flexibility of internal layout and space allocation to workers. Within a house, rooms may have barriers removed between them.

To value the flexibility that open plan configurations give can be outlined in general terms, but each case requires its own specific numbers because of the many possibilities for change that exist. It is not possible to say a priori in general terms whether open plan is viable or not, but rather only on an application-by-application basis. The A versus NA comparative analysis below shows how the value of open plan design can be established.

5.2.4 The Market

In principle, the market should place a higher price on a building that has in-built adaptability features. It has value to a buyer, but how that value compares with the value calculated using Eq. (1.1) would be unknown in any negotiation between seller and buyer.

5.2.5 Costing Uncertainty

Commonly, new building work might be undertaken on a lump sum basis, subject to adjustments, which are possible in most contracts. Renovation and addition/extension work might preferably be done by builders on a cost reimbursable basis, because of the unknowns in the building being altered. Cost reimbursable work and final building cost could be anticipated to have greater variability or uncertainty than lump sum work and final cost. An examination of Eq. (1.1) shows that the option value increases with uncertainty. The more uncertain are the costs and disbenefits at time of adaptation, the higher the adaptability value. Methods for estimating costs and disbenefits in the presence of uncertainty are outlined in Chap. 2.

5.3 Background

Building conversions may be with respect to interiors, exteriors, building volume (or space) and use. The literature on this is predominantly with respect to established buildings, designed and built without forethought to alteration (NA form). By contrast this chapter scrutinises deliberate built-in flexibility (A form) [4, 6].

5.3.1 Typical Alteration—NA Form

Changing interiors includes reconfiguring space layout from closed individual spaces to open-plan spaces, or vice versa. This may have as its aim to enhance space usage efficiency and reusability of spaces, or to take advantage of existing space potential to accommodate minor shifts in needs. Common practice involves a reconfiguration of rooms by relocation of interior partitions, with associated modification in ceiling and floor finishes.

Changing volume or space may be in response to an increase or reduction in demand. Building expansion might occur either vertically or horizontally. Building volume contraction might be carried out, for example, through dividing and possibly subletting a part of the building space.

Examples of *altered use*, which might be referred to as 'adaptive reuse', include: refurbishing heritage buildings following obsolescence in terms of their original use; and converting redundant offices into apartments, perhaps because of an oversupply of office space, shifting technology or shifting demands. However, altered use is less relevant to houses than commercial buildings, except for the incorporation of new technology, or shift in usability due to a life transition of the owner. Examples of this include making houses accessible throughout and more readily usable for aged people, particularly in bathroom and kitchen areas.

5.3.2 Flexibility: Built-in (A) Form

In comparison to the NA form, there is much less literature on the A form.

There is a belief, but generally not supported with analysis, that it is inefficient to construct buildings with a single use in mind; rather, there is a need to design buildings with the ability to change if necessary. There is qualitative discussion in the literature on the trade-off between possible up-front extra costs and long-term benefits, and the architectural and engineering aspects of flexible design. The literature: talks of open buildings, where the façade is the only thing that changes, and easily accessible building services; and suggests that designing infrastructure with the foresight for change will not cost much more in original outlay. A consensus view is that the total building cost would only increase by at most a few percent, and this is regarded as insignificant when considering that the building may be obsolete if demand shifts.

The notion of built-in flexibility overlaps with that of the 'universal design' concept, where a residential-style building is designed and built as friendly to all persons, including the disabled, young and old. The universal design concept addresses the shifting needs of building inhabitants. However, unlike this chapter's view on flexibility, it does not have a future time at which the design is changed; rather the design stays constant over the life of the dwelling. The literature suggests that such designs will only add small extra costs, and could benefit in terms of longevity of owner's usage, and expanding the tenant pool if applied to rentable buildings.

Alterability might be discussed in terms of a number of approaches, which can be achieved through changes in design—space layout, space use and volume. Such design is suggested to lead to longer service lives because changes can be achieved at lower cost.

5.4 Comparative Analysis

5.4.1 Outline

The analysis here is given in general terms, which can be applied to any deliberate built-in flexibility situation. Later, some specific numerical values are given to the analysis.

Two forms of alteration (A—deliberate built-in flexibility that facilitates alteration; and NA—no built-in flexibility, the more usual practice) are considered and compared in financial, environmental and social terms, singly and in combination. The analysis considers a building (constructed at time $t = 0$), with the potential to be altered, at some time $t = T$ in the future. There is a trade-off between financial, environmental and social impacts now and those in the future. The alteration, however, need never occur, depending on the building owner's situation in the future. That is, the alteration is discretionary or an option. The time, T, can be varied in the calculations in order to show the relationship between time of any alteration and flexibility value.

5.4.2 Financial Analysis

Existing publications either do not put any financial or quantitative value on deliberate built-in flexibility, or use traditional financial market options analogies, which have been criticized. Conventional discounted cash flow analysis and decision trees are not applicable because of the uncertainties present and because the alteration may or may not take place at the discretion of the building owner. Here the financial value is established through this book's rational analysis based on cash flows [3, 5].

To establish the option value (but not the complete financial viability), only costs and disbenefits at T are taken into account (Fig. 5.1). (Disbenefits can be treated similarly to costs. Positive benefits are assumed to be the same for both the A and NA forms.) Expected values, E[], and variances, Var[], of all costs and disbenefits for both A and NA forms at T are estimated, or if they occur beyond T, are discounted to T. There are a number of ways by which estimates may be obtained as mentioned in Chap. 2. Because estimates for the A and NA forms are based on similar assumptions, it could be anticipated that there would be very strong correlation between the estimates for the A and NA forms.

To ascertain the value of flexibility (A form) compared with more conventional (business-as-usual—BAU) practice (NA form), the difference between the NA and A forms is examined. Figure 5.1 applies. The net costs and disbenefits at T, X_T, is defined as in Appendix 2.11.1, where $i = T$. In a comparative analysis, because the difference between the NA form and the A form is being looked at, costs and disbenefits, Y_{Tk}, $k = 1, 2, 3, \ldots, \mu$, of the NA form are considered positive, while costs and disbenefits, Y_{Tk}, $k = \mu + 1, \mu + 2, \ldots, m$, of the A form are considered

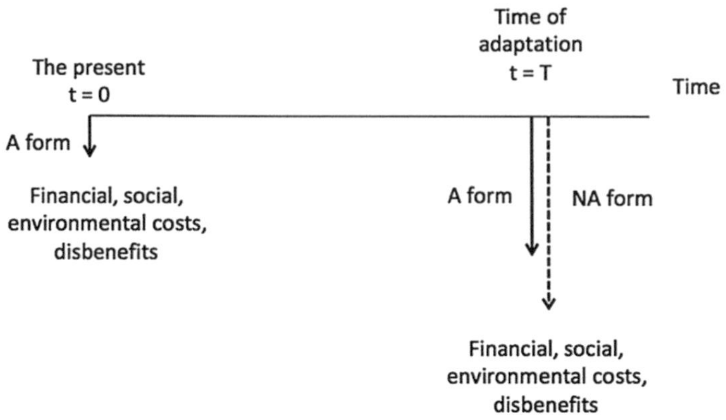

Fig. 5.1 Schematic of the valuation inputs

negative, and the X_T is the difference between the NA and A forms. Strict accounting conventions are not followed. Rather, everything is looked at from the viewpoint of the investor or option holder, such that something favourable to the investor (avoiding the costs and disbenefits of the NA form at time T) is taken as positive and something unfavourable to the investor (having the costs and disbenefits of the A form at time T) is taken as negative.

Present worth then follows from Appendices 2.11.2 or 2.11.3. And the flexibility value is given by Eq. (1.1).

This flexibility value is then compared with the additional costs and disbenefits (equivalently the premium) flowing from building in flexibility at time 0. Financial viability is established for flexibility when the flexibility value exceeds the collective additional initial costs and initial disbenefits. However, where an owner has restricted access to finance at time 0, or wishes the cheapest initial building cost, this may constrain the owner from investing in flexibility.

5.4.3 Environmental Analysis

Environmental issues have recently started to be valued quantitatively, with cost comparisons between new construction, extensive renovation, refurbishment and continued operation of residential buildings. Here, life cycle assessment (LCA) analysis is preferred as being the most applicable approach. The analysis looks at the environmental impacts at times $t = 0$ and T, assuming impacts between $t = 0$ and $t = T$ and subsequent to T to be the same for both A and NA forms. Should these assumptions not be so in particular cases, the differences can be included in the analysis.

Environmental issues involved in deconstruction and construction (at t = 0 and/or t = T) include:

- *Resource consumption*: materials and energy use;
- *Emissions*: equipment-producing emissions; embodied emissions within materials used or removed; and
- *Waste generation*: material waste in deconstruction and construction; material reuse.

This list is not exhaustive, and would be tailored to each specific case. When compared to full building demolition, the contribution of flexibility to sustainability through saved resources and energy is acknowledged, among other instances, through the wide-held belief that the greenest buildings are the ones we already have (ignoring operation and maintenance issues).

A well-designed A form building leads to the reuse of materials and a subsequent reduction in the environmental impact in terms of waste, materials consumption, embodied energy and emissions. A well-designed A form building also leads to lesser construction site time and effort in the alteration. The flow-on is reduced construction energy use and emissions, and nuisance related dust and noise.

5.4.4 Social Analysis

Social issues have recently started to be valued quantitatively. Here, social life cycle assessment (SLCA) analysis is used. SLCA follows the approach of environmental LCA but in assessing social issues, except for those issues such as people flow within the building, and aesthetics, which might be analysed based on surveys of stakeholders' opinions.

Social issues, associated with altering buildings, result from physical and/or psychological impacts on stakeholders. Those social issues connected with deconstruction and construction reflect the quantity and type of physical work involved in the alteration. The A and NA forms require differing extents and type of work activities at t = 0 and T. This, in turn, leads to differing extents of the social issues. Some relevant social issues (at t = 0 and/or t = T) include:

- *Workers*: health and safety in construction; employment opportunities;
- *Neighbours*: disruption to traffic in the immediate neighbourhood; emitted pollutants, dust, noise and vibration; and
- *Owners*: level of comfort; identity due to long-term inhabitancy; inconvenience due to compulsory move-out.

This list is not exhaustive, and would be tailored to each specific case. The potential for increased accidents (and hence lower safety), and greater disturbance and inconvenience to the owner, neighbours and local traffic flow, is associated with increased construction site-based activities. However, shorter site times could be anticipated to lead to lesser total employment hours.

5.5 Numerical Studies

5.5.1 Outline

Numerical studies are given for a range of differences between the NA and A forms, time of adaptation, interest rate and uncertainty in the future estimates for the NA and A forms.

Let Y_{T1} be the (collective) NA form costs and disbenefits at T, and Y_{T2} the (collective) A form costs and disbenefits (Fig. 5.1),

$$X_T = Y_{T1} - Y_{T2}$$

The following ranges in variables are considered:

$$Y_{T1} - Y_{T2} = \alpha Y_{01},\ \text{with the multiplier}\ \alpha : 0.5,\ 1.0,\ 2.0,\ 5.0$$
$$\text{Time to adaptation (years), T}: 2, 5, 10, 15, 20$$
$$\text{Interest rate (per annum), r}: 0.05, 0.1, 0.15$$
$$\text{Coefficient of variation of}\ Y_{TI} - Y_{T2},\ \text{COV}: 0.2, 0.5, 1.0$$

Y_{01} refers to the collective costs and disbenefits of the A form at time 0. The base values for the sensitivity-style calculations carried out here correspond with $\alpha = 2.0$, $T = 10$ years, $r = 0.1$ per annum, and $COV = 0.5$.

In Figs. 5.2, 5.3, 5.4 and 5.5, adaptability viability is plotted against these variables. Viability is here defined as the difference between the adaptability value (calculated from Eq. (1.1)) and the collective costs and disbenefits at $t = 0$ necessary to achieve adaptability (A form). In Figs. 5.2, 5.3, 5.4 and 5.5 it is the trends, which are important, rather than the absolute values, because collective costs and disbenefits have been unitised. Figure 5.2 shows, as anticipated, that the viability increases with the difference between the collective adaptation costs and disbenefits of the NA form and the collective adaptation costs and disbenefits of the A form. Figure 5.3 shows

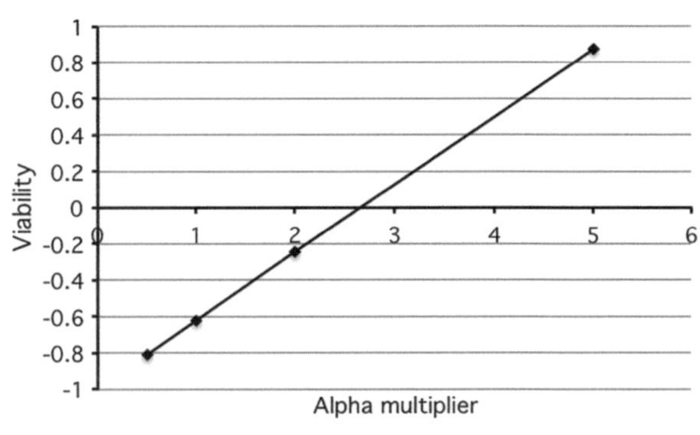

Fig. 5.2 Viability of adaptability versus difference in adaptation costs and disbenefits at T

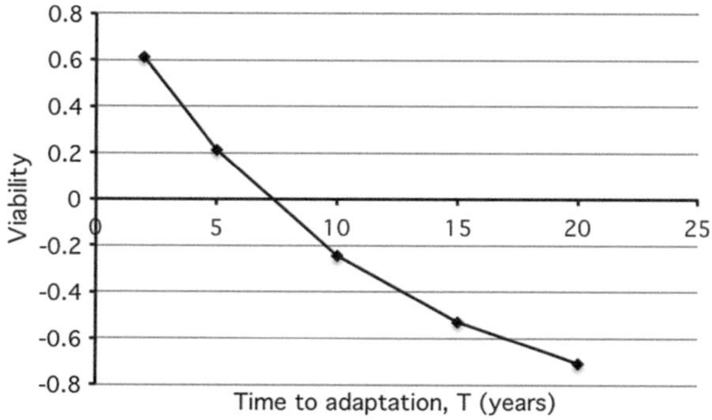

Fig. 5.3 Viability of adaptability versus time to adaptation

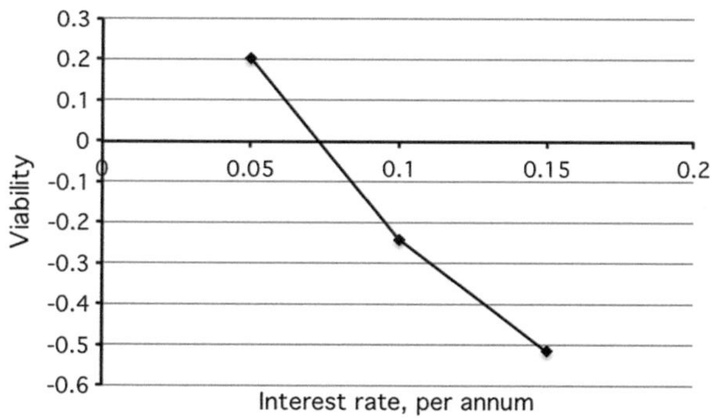

Fig. 5.4 Viability of adaptability versus interest rate

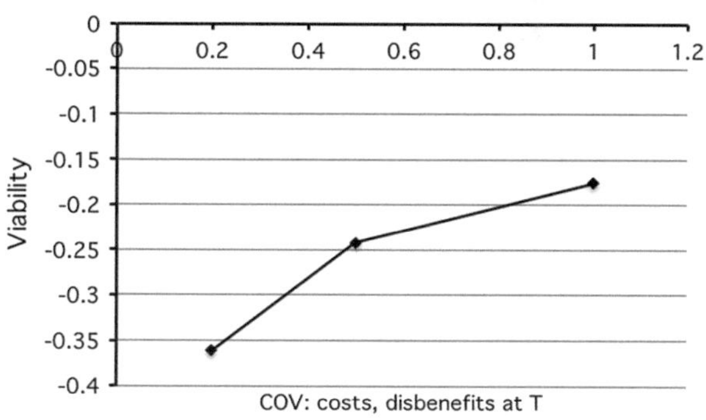

Fig. 5.5 Viability of adaptability versus future uncertainty

the decline in viability the further into the future adaptation takes place. Figure 5.4 shows the decline in viability with interest rate. Figure 5.5 shows the increase in viability with uncertainty in the collective cost and disbenefit estimates at time t = T.

5.5.2 *Sustainability*

For A form design well-thought out, it can be demonstrated that the built-in A form is generally better than the non built-in NA form according to financial, environmental and social impacts. Accordingly, the A form would be considered a more sustainable alternative, irrespective of how the environmental and social impacts are weighted relatively or dealt with collectively [1].

Environmental and social criteria enhance the financial value of deliberate built-in flexibility in buildings.

The sustainability argument in favour of the deliberately built-in A form over the non built-in NA form is not strongly influenced by whether the analysis is carried out from the owner's viewpoint or the community's viewpoint. However, the owner's decision making could be anticipated to be strongly influenced by personal interest and personal values.

5.6 Summary

Viability for built-in adaptability cannot be stated in general, but requires an examination of each situation. However, some general conclusions from the author's studies on designed-in/built-in adaptability are:

- Building in adaptability/flexibility involves some extra initial costs and disbenefits, which can be minimised through good design and engineering. However, this may deter those who want cheapest initial cost.
- Building in adaptability is most attractive where adaptation is anticipated in the shorter term. The attractiveness of longer-term adaptation will depend on interest rates used and any initial costs and disbenefits. For public sector buildings, using environmental/social discounting will make longer term adaptation more viable.
- Built-in adaptability will be more viable for the public sector than the private sector because of the lower rates and intangibles (social and environmental) used in feasibility studies.
- Social and environmental issues add to the value of building in adaptability, over pure financial arguments. In coming years, people may value social and environmental issues more, thus further enhancing the case for built-in adaptability.
- The 'ingenuity' of designers and engineers can overcome hurdles where there is no forethought to adaptability.
- The main arguments in favour of building in adaptability may be aesthetics and function.

5.7 Discussion

Conclusions based on numerical studies will necessarily be based on certain design assumptions, style of building and material choice, and a given country. Accordingly, the methodology, rather than the absolute numbers, is emphasised in this chapter. The numbers given in the numerical studies are indicative, but will differ with reader assumptions and geographical locations. Different forms of built-in flexibility are possible, and with building practices location-dependent, there are too many variables to enable any generalisation. The methodology given in this chapter is general and it carries over to other situations and locations, but the actual numbers and building practices will not.

Deliberately building in flexibility can typically be shown to be financially viable up to time periods of about a decade, but not longer time periods unless the interest rates are low. Social and environmental impacts increase the viability. However, it is unclear as to how building owners will interpret the favourable financial, social and environmental impacts and what unstated decision making processes building owners go through.

The resale value of a building containing deliberate built-in flexibility should in principle be greater than a building without such a feature. However, it is unclear as to what value the market would place on this; to a buyer considering a possible future alteration, its worth follows Figs. 5.2, 5.3, 5.4 and 5.5, but in any negotiation between buyer and seller this information would be unknown to the seller.

Examining the background to Eq. (1.1), the deliberate built-in flexibility financial value increases with uncertainty about future costs and disbenefits. The predominant source of this uncertainty lies in the building work method adopted and the scope of the work, with lesser uncertainty due to material choice. The building work might typically be undertaken on a cost reimbursable basis because an existing building is being modified and because of the poor scope definition, rather than on a fixed price basis as is more common with new buildings, which have good scope definition. The final outcome cost for cost reimbursable work is never definite.

The uncertainty in the cost and disbenefit estimates could be assumed to grow with time, on the basis that estimates further into the future are less definite. The calculations show, however, that this uncertainty has to be reasonably large, relative to expected cost and expected disbenefit value, before it starts to alter the broad conclusions in the case example. Similarly, the value placed on social and environmental impacts could be anticipated to increase over time, as people readjust their value systems in line with increasing public sustainability imperatives.

5.8 Closure

This chapter showed a method for valuing deliberately built-in flexibility in buildings. This was exampled in numerical studies. The built-in flexibility form was compared with the non built-in flexibility form. Financial viability for built-in flexibility can be demonstrated for smaller interest rates and smaller times to alteration, but not for larger interest rates and/or time to alteration. The numerical studies made certain assumptions, and altering these assumptions will give different calculated values; however the methodology remains the same.

Whether building in flexibility is viable or not, from the building owner's financial viewpoint, cannot be said in general terms, but rather requires an individual analysis for each situation. No general conclusions on the viability can be drawn, but rather depend on the specifics of each situation. In some situations, deliberate built-in flexibility will be worthwhile, while in others it may not. Intangibles and sustainability arguments increase the viability of building in flexibility but at the present time, based on numerous observations by the author, it is believed that few building owners would apply much weighting to these intangibles and sustainability compared to the financial aspects.

The architecture literature suggests that there is value in flexibility, but has stopped short in quantifying this value. This chapter provides an analysis that gives the value of any flexible building infrastructure. The approach will be useful to the construction industry and building owners contemplating prolonging the useful life of buildings through alteration over time. The method given in this book has broad applicability, and extends to flexibility analysis generally, for example to conversion of buildings generally to alternative usages, and even including urban 'planning' changes.

5.9 Extensions

The analysis given in this chapter can be applied to any infrastructure type—low or high rise buildings, commercial or residential buildings or civil infrastructure.

Fine tuning of deliberately built-in flexibility features is possible, such that built-in flexibility might become more commonplace. The analysis given above will remain the same, but the numerical values for different applications will not. With creative thought, built-in flexibility could become the norm.

In this chapter it was stated that creative design is involved in how flexibility might be incorporated in order to produce the best outcome for the owner. Future research might look at how creative design might be captured in the analysis.

It is unclear whether a building with built-in flexibility would be more attractive to the market, and whether the market would pay more for such a building. A study on this could clarify the situation.

For houses, it is anticipated that different people would have different perceived utility of built-in flexibility, depending for example on whether a person is attached to a location or a house, or not. A person's age would also affect this utility, with younger people being more mobile as their workplace locations move, and older people thinking of family stability. The utility of built-in flexibility could be researched.

Specific reasons why people modify houses and what is modified are not explored in the chapter, nor are the socio-economic backgrounds of the people and their conversions. Societal and demographic aspects may hinder or support the capacity to do alterations. There may be unfavourable issues that prevent alterations or limit the capacity to do alterations. This could be the subject of future research.

The social and environmental issues represent intangibles. An attempt could be made to establish values that people put on these intangibles. Generally, it is anticipated that intangibles will increase the viability of designing in flexibility features.

For houses, the biggest argument in favour of having pre-thought flexibility may be an aesthetic one. The NA form always looks like a house altered or converted, while the A form can seamlessly change the existing building. The aesthetic value of the A form over the NA form needs exploring. Similarly, function and people movement within houses of A and NA forms need exploring.

An alternative view is to regard the choice at $t = 0$ between the deliberate built-in A form and the NA form as an exercise in risk management. Here risk is used in the sense of being a function of outcome likelihood and outcome magnitude [2]. This approach would require building owner estimates of future probabilities that an alternative building will be required, and probabilities associated with the timing of the future alteration, or average community statistics could be used. The applicability of risk management thinking to building owners could be explored.

It is recognised that building practices vary around the world, and while the methodology of the chapter will apply to all building types, the conclusions are location-specific, and differences in building types and practices could be explored.

References

1. Carmichael DG (2013) Problem solving for engineers. CRC Press, Taylor & Francis, London
2. Carmichael DG (2016) Risk–a commentary. Civ Eng Environ Syst 33(3):177–198
3. Carmichael DG (2016) A cash flow view of real options. Eng Econ 61(4):265–288
4. Carmichael DG (2018) The case for building in adaptability in houses. Joint Asia-Pacific Network for Housing Research (APNHR) and Australasian Housing Researchers (AHR) Conference, Griffith University Cities Research Institute, Gold Coast, 6–8 June 2018
5. Carmichael DG, Hersh AM, Parasu P (2011) Real options estimate using probabilistic present worth analysis. Eng Econ 56(4):295–320
6. Carmichael DG, Taheriattar (2018) Valuing deliberate built-in flexibility in houses-exampled. Int J Strateg Prop Manag 22(6):479–488

Chapter 6
Project Delivery—PPP Guarantees

6.1 Introduction

A public-private partnership (PPP, P3) may be classified as belonging to what are generally referred to as concessional delivery methods, which also incorporate privately-financed initiatives/projects (PFI, PFP), and different build-own-operate-transfer (BOOT) varieties [2]. PPP delivery is popular with the public sector because it enables infrastructure to be designed, constructed and operated using private funding. It can also be used by the private sector, for example coal-washing facilities on mine sites, though the majority of applications appear to be with the public sector. The relevant public sector authority (referred to below as 'the authority') uses the finance and skills of a private sector consortium (referred to below as 'the concessionaire') in this delivery. In return, the concessionaire is given time (a concession period) over which its investment can be recovered.

The following is written for public-private partnership (PPP, P3) toll road projects, but it applies generally to all PPP type projects. The uncertainty, risk and fairness in PPP agreements in toll road projects, specifically the financial aspects of such agreements, may be addressed in part by having options. Within the context of financial agreements and toll roads, the option for example may translate to adjusting future revenue in response to uncertain and shifting future road demand. The option right may come about in return for some direct or indirect cost (premium) to the option holder.

Depending on how a PPP agreement is structured, the concessionaire carries differing degrees of financial risk, primarily arising from road usage, patronage or demand uncertainties. While capital costs and ongoing operation costs are reasonably predictable, demand is not, and is influenced by the magnitude of the tolls being charged, travel times, vehicle operating cost, and the availability of alternative roads and transport. The viability analysis of the project from the concessionaire's viewpoint, among other things, looks at the financial risk carried, and attempts to reduce

D. G. Carmichael, *Future-proofing—Valuing Adaptability, Flexibility, Convertibility and Options*, Management in the Built Environment,
https://doi.org/10.1007/978-981-15-0723-6_6

this risk through adjusting the agreement between the parties. The concessionaire may request subsidies or guarantees from the authority. The authority may, in turn, request reciprocal guarantees. An agreement involving guarantees, if properly structured, allows the risk to each party to be managed, and make the project more viable. If the risk being carried is considered unacceptable, either party might withdraw from the project.

The existing literature focuses on revenue-related guarantees introduced to deal with uncertainties in demand during the operational phase. These guarantees can take different forms. Third parties, such as insurance companies, may be involved, but the intent is the same whether third parties are involved or not, namely to assist the authority and the concessionaire with viability and risk management. Such guarantees may be valued using an options analysis. In general, the revenue is uncertain, being based on road usage, patronage or demand. This uncertainty needs to be captured in any analysis. Typically, the literature uses traditional financial market options techniques, and applies these by analogy. Each option is presented and analysed in stand-alone papers, and relies on the high level of mathematical skills of their authors. In contrast, this chapter presents a single unifying approach for analysing all PPP toll road options.

The chapter is structured as follows. The literature on PPP toll road options is first reviewed. The book's unifying approach is then presented. Each existing proposed PPP toll road option is presented and interpreted in terms of this book's approach. Discussion and conclusions follow. The chapter is written in terms of two parties to the PPP agreement, namely the authority and the concessionaire.

It is emphasized that only options dealing with financial aspects are dealt with. Options of a physical nature (although involving money), for example, in terms of increasing the number of lanes of a road or delaying the construction of a road are addressed in Chap. 4.

The chapter provides an original and unified approach to PPP toll road options. The chapter will be of interest to anyone involved in PPP toll road projects.

6.2 Background

PPP financial agreements between the authority and the concessionaire can involve a range of guarantees or adjustments under differing names or descriptors:

- Minimum revenue guarantee (MRG). The guarantee involves the authority paying the concessionaire if the actual toll revenue falls below a pre-agreed threshold. This puts a limit on the revenue downside for the concessionaire.
- Buyout. The authority holds the right to buy the concession back before the end of the concession period, at a predetermined exercise price, subject to certain conditions.
- Revenue-sharing. The authority holds the right to claim a percentage share of excess revenue when the revenue exceeds an agreed upper limit or threshold.

- Restrictive competition guarantee. This guarantee secures a road's revenue against loss caused by competing roads.
- Collar. A collar combines both lower and upper revenue thresholds to create a band. The concessionaire holds an option on low revenue. The authority holds an option on high revenue.
- Traffic floor and ceiling (TFC). Traffic floor and ceiling is based on pre-agreed lower (floor) and upper (ceiling) traffic levels. The concessionaire holds the traffic floor option, while the authority holds the traffic ceiling option. It is the same as a collar, but with two differences—an up-front cost or premium, and a guarantee covering only part of any traffic shortfall or traffic exceedance (referred to as a partial coverage guarantee).
- Toll adjustment mechanism (TAM). This is similar in intent to MRG (a guaranteed minimum revenue to the concessionaire), however TAM gives the concessionaire the right to adjust tolls to achieve a desired revenue.

Within existing publications, traffic or revenue is commonly assumed to follow a time series such as geometric Brownian motion with associated volatility measure. Monte Carlo simulation may be used to generate realizations, or the Black-Scholes equation might be used if applicable. By contrast, this book's approach does not have restrictive assumptions on time series or volatility, nor does it require exercise prices to be deterministic or at a single point in time; uncertainty is incorporated through variance estimates. There is also no need to distinguish between option types, for example a call (equivalent to a purchase) option or put (equivalent to a sale) option [2, 3], rather each case considers the cash flows from the viewpoint of whoever holds the option.

6.3 Cash Flows

Typically, for road PPPs, the cash flows involving an option can be thought of in terms of a cash outflow and a cash inflow at time $i = T$, the time of exercising the option (Fig. 6.1). $1 \leq T \leq n$, where n is the concession period. In Fig. 6.1, Y_{T1} and Y_{T2} refer to cash inflow and cash outflow respectively at time T, such that, with respect to Appendix 2.11.1, $X_T = Y_{T1} - Y_{T2}$. Y_{T1} and Y_{T2} may be deterministic or probabilistic, and are from the viewpoint of the option holder. The origins of these cash flows differ in each application, and are explained below. In some applications a cash flow may be revenue foregone, while in other applications, the two cash flows may represent the cash flows associated with exercising and not exercising an option.

Where cash flows connected to the option extend beyond T, that is, over time periods $i = T + 1, T + 2, \ldots, n$, then these are collectively discounted to time T, such that

$$E[Y_{T1}] = \sum_{i=T+1}^{n} \frac{E[Y_{i1}]}{(1+r)^{i-T}}$$

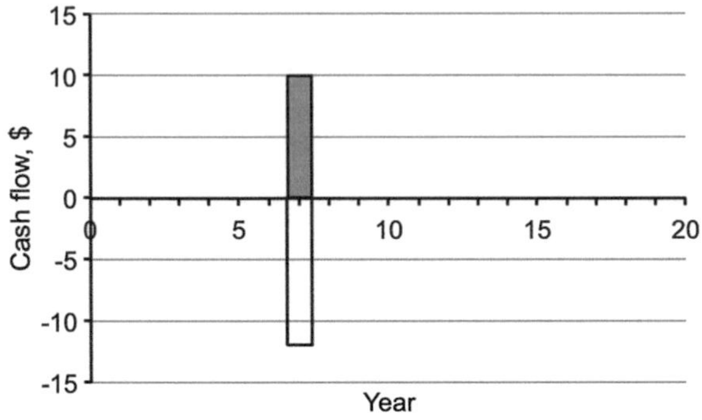

Fig. 6.1 Example cash flows involved in an option at time T

$$\mathrm{Var}[Y_{T1}] = \sum_{i=T+1}^{n} \frac{\mathrm{Var}[Y_{i1}]}{(1+r)^{2(i-T)}} + 2\sum_{i=T+1}^{n-1}\sum_{j=i+1}^{n} \frac{\mathrm{Cov}[Y_{i1}, Y_{j1}]}{(1+r)^{i+j-2T}}$$

Similar expressions apply for Y_{T2}. In this sense, Y_{T1} and Y_{T2} become an 'equivalent' cash inflow and an 'equivalent' cash outflow, respectively, in year T.

The content of Appendix 2.11.2 applies. The option value follows from Eq. (1.1).

6.4 Existing PPP Road Options

6.4.1 Outline

Existing PPP toll road option cases are grouped according to the following descriptors: Minimum revenue guarantee (MRG); Buyout; Revenue-sharing; Restrictive competition guarantee; Collar; Traffic floor and ceiling (TFC); and Toll adjustment mechanism (TAM). In each case, this book's analysis is compared with the existing literature. To do this comparison, assumptions are made as compatible as possible with the existing literature, but necessarily not exactly the same, primarily because this book's approach uses variance instead of volatility, and there is no universal agreement as to what volatility should be used or to the conversion between volatility and variance [8]. Appendix 2.11.4 gives the result used to guide the conversion between variance and volatility. In the general case, the daily traffic, toll and any thresholds or equivalent could be anticipated to vary over time. The form of the analysis presented in this chapter remains the same, should any of these vary from year to year or be constant from year to year [4].

6.4.2 Specific Notation

The following gives the particular toll road notation used.
Revenue related

C_i	maximum revenue guarantee (threshold) in year i (pre-agreed); a revenue cap; related to TC_i where a traffic ceiling is defined
f	revenue cap growth rate (constant cap growth per year)
f_c	minimum threshold growth rate (percent/year, compounding yearly)
F_i	minimum revenue guarantee (threshold) in year i (pre-agreed); a revenue floor; related to TF_i where a traffic floor is defined
g	revenue growth rate (constant revenue increase per year)
g_c	revenue growth rate (percent/year, compounding yearly)
R_i	revenue in year i
α, β	percentages, or fractions $0 \le \alpha, \beta \le 1$

Traffic related

TC_i	traffic ceiling guarantee in year i, $TC_i = Tcr \times E[v_i]$
Tcr	traffic ceiling ratio, used in establishing the traffic ceiling guarantee TC_i
TF_i	traffic floor guarantee in year i, $TF_i = Tfr \times E[v_i]$
Tfr	traffic floor ratio, used in establishing the traffic floor guarantee TF_i
$Toll_i$	toll per vehicle in year i (possibly, pre-agreed)
$Tollcap_i$	toll cap in year i (pre-agreed)
v_i	traffic (vehicles/year) in year i
γ	traffic growth rate (constant number of vehicles per year)
γ_c	traffic growth rate (percent/year, compounding yearly)
ϕ_c	toll growth rate (percent/year, compounding yearly)

General

c, a	superscripts denoting concessionaire and authority, respectively
wi, wo	superscripts denoting with-invoking and without-invoking TAM, respectively

Where the subscript i is omitted, then the variable takes a constant value for all i.
Currencies. USD US dollar, \$; HKD Hong Kong dollar (HKD1.00 $\approx$ USD0.13); CNY Chinese yuan (CNY1.00 $\approx$ USD0.16); INR Indian rupee (INR1.00 $\approx$ USD0.016)

6.4.3 Minimum Revenue Guarantee

Minimum revenue guarantee (MRG) refers to a mechanism for limiting the revenue downside to the concessionaire, resulting from revenue uncertainty. The authority provides a guarantee of a minimum annual revenue (a threshold value), F_i, to the concessionaire. This can be viewed in terms of the concessionaire holding an option in each year of the concession period. The concessionaire exercises each option, and claims the revenue shortfall, when the actual annual revenue is lower than this defined minimum threshold. The total value of having the yearly options is the sum of the yearly option values.

In any year $i = T$ (Fig. 6.1), the option is only exercised if the revenue shortfall, $X_T > 0$, where: Y_{T1} = the minimum guarantee (threshold) value (F_i at $i = T$); and Y_{T2} = the revenue (R_i at $i = T$).

Example. Adapting Brandao and Saraiva [1], using a comparable traffic volume standard deviation equal to 30% of its expected value: concession period, $n = 25$ years; interest rate, $r = 15\%$ per annum; minimum guarantee level (constant), $E[F_i] = \$1.5B$, $Var[F_i] = 0$; revenue (constant), $E[R_i] = \$1.9B$, $Var[R_i] = (\$0.57B)^2$. Here, F_i is 80% of $E[R_i]$. F_i and R_i are assumed to repeat for all $i = 1, 2, \ldots, n$. For $i = T$ (using Appendix 2.11.2), $E[X_T] = -\$0.4B$, and $Var[X_T] = (\$0.57B)^2$. From this, E[PW] and Var[PW] can be obtained and the option value for any year calculated (Fig. 6.2—solid curve).

Comparison with the literature. The sum of the yearly option values for 25 years is approximately \$0.52B. This is higher by approximately 0.4%, as a proportion of the threshold value, when compared with Brandao and Saraiva [1]. The yearly option values vary with the level of guarantee as shown in Fig. 6.2. The level of guarantee, that constitutes the minimum threshold in year i, is defined as a percentage of the expected annual revenue in year i, namely $E[R_i]$. The yearly option values increase as the level of guarantee increases, as anticipated, and this trend agrees with the observations of Brandao and Saraiva [1].

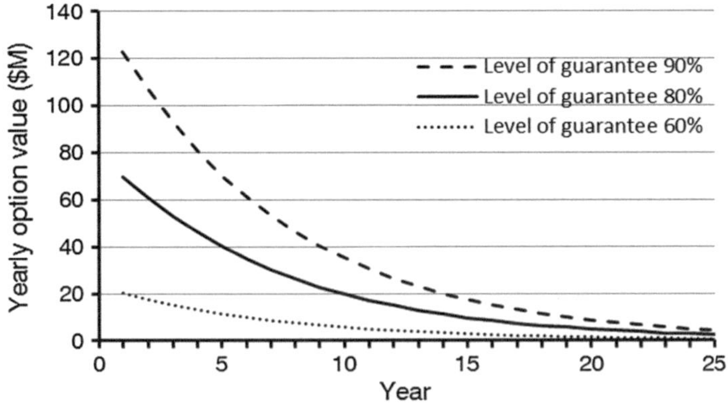

Fig. 6.2 Minimum revenue guarantee example; yearly option value versus time for different levels of guarantee

6.4.4 Buyout

A buyout option gives the authority the right to buy the concession back before the end of the concession period, at a predetermined buyout cost (exercise price), subject to certain conditions. The buyout option may be exercised, at time T, only if the updated value of the revenue remaining till the end of the concession period (that is, over the period from $i = T$ to $i = n$) exceeds a pre-agreed level (equivalently, a buyout cost). The updated value is the (updated) present worth (discounted to T) of the remaining revenue cash flows. Expressed differently, this means taking all the cash flows from $i = T$ to $i = n$ and discounting them to time T to give a collective discounted value—the updated value. The exercising may be defined to occur at a pre-specified year within the concession period, or in any year of the concession period.

In any year $i = T$ (Fig. 6.1), the option is exercised if $X_T > 0$, where: Y_{T1} = the revenue (R_i) for years $i = T, T + 1, T + 2, \ldots, n$ discounted to year T; and Y_{T2} = the buyout cost (F_i at $i = T$).

Example. Consider the buyout option example of Power et al. [10], using a comparable traffic volume standard deviation equal to 25% of its expected value: traffic (million vehicles) in year 1, $E[v_1] = 36.085$; $Var[v_1] = 9.021^2$; toll per vehicle (constant), Toll = \$10; traffic growth (million vehicles per year), $E[\gamma] = 5$; $Var[\gamma] = 1^2$; buyout cost at year 6 (buyout multiplier of 1.5), F_6 = \$8,600M; interest rate, r = 11.6% per annum; concession period, n = 25 years. The buyout cost (here a constant) equals the product of a buyout multiplier (here, 1.5) and the present worth (at year 0) of revenue over the 25-year concession period (here, \$5,700M). The revenue in year i follows from Results (I) and (II) of Appendix 6.8, with $E[R_1]$ = \$361M and $Var[R_1]$ = (\$90M)^2.

Consider exercising the buyout option in year 6, interpreted from the viewpoint of the option holder (the authority). The buyout cost (a known amount), $E[Y_{62}]$ = F_6 = \$8,600M, and $Var[Y_{62}] = 0$ based on Power et al. [10]. In calculating Y_{61} according to Appendix 6.8, the R_i, $i = 6, 7, 8, \ldots, 25$, could be assumed to be strongly correlated. Discounting, E[PW] = –\$399M; Var[PW] = (\$541M)^2. From Eq. (1.1), OV = \$72.5M.

Figure 6.3 (solid curve) shows how the option value varies with year of exercising, assuming that the buyout cost remains the same at \$8,600M. The current set of values might be considered favourable to the authority over the concessionaire in terms of early buy out, and would need negotiation, and in particular negotiation perhaps on a variable buyout cost. Figure 6.3 shows the influence of altering the value of the buyout multiplier through using example buyout multipliers of 1.5 (as in the numerical example above) and a slightly larger 1.6. The buyout cost equals the product of the buyout multiplier and the present worth (at year 0) of revenue over the 25-year concession period.

With a buyout cost, F_i, defined as a multiplier of the present worth of the revenue remaining till the end of the concession period (buyout cost reducing with time),

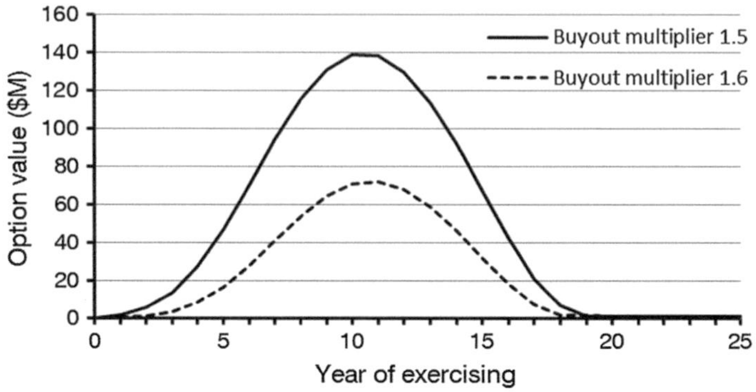

Fig. 6.3 Buyout example (constant buyout cost); option value versus time for different levels of buyout

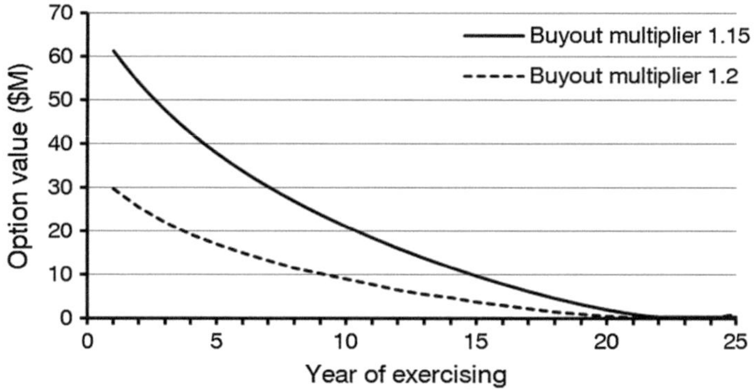

Fig. 6.4 Buyout example (reducing buyout cost); option value versus time for different levels of buyout

Fig. 6.4 shows how the option value varies with year of exercising and buyout multiplier magnitude. This result is similar to revenue-sharing. Whereas revenue-sharing allows the authority to collect excess revenue if revenue is higher than an upper threshold, buyout gives the authority the right to buy back the project, and collect the full revenue for the remainder of the concession period.

Comparison with the literature. This book's approach gives an option value (based on exercising in year 6, with a fixed buyout cost) higher by approximately 0.5% as a proportion of the buyout cost, when compared with Power et al. [10].

Figure 6.3 shows that the optimal value of the buyout option occurs near the middle of the concession period, where the plot peaks. This reflects the increasing yearly revenue with time, countered by the decreasing present worth of future revenue with time. This optimality finding agrees with the conclusion of Power et al. [10]. The option value varies with year of exercising and decreases with increasing level of buyout cost for both constant and varied buyout cost. These trends agree with the observations of Power et al. [10].

6.4.5 Revenue-Sharing

A revenue-sharing option gives the authority (as option holder) the right to claim a percentage share, β, of the revenue that exceeds an upper limit (maximum revenue cap) in any year. The cap is adjusted upwards by a constant amount each year. The concessionaire retains a $(1 - \beta)$ share of this excess revenue. The option is exercised in any year when the revenue exceeds the pre-agreed limit. The total value of having the yearly options is the sum of the yearly option values.

In any year $i = T$ (Fig. 6.1), the option is exercised if the excess revenue above the cap, $X_T > 0$, where: Y_{T1} = the revenue (R_i at $i = T$); and Y_{T2} = the maximum revenue cap (C_i at $i = T$). The authority receives β percent of the revenue excess, that is, βX_T. The option value is calculated based on the present worth derived from an expected value $\beta E[X_T]$, and a variance $\beta^2 Var[X_T]$. (This follows from the results in Appendix 6.8.)

Example. Consider the example of Song et al. [12]: revenue in year 1, $E[R_1]$ = CNY138M; $Var[R_1]$ = (CNY48M)^2; revenue growth per year, $E[g]$ = CNY20M; $Var[g]$ = (CNY8M)^2; revenue cap in year 1, $E[C_1]$ = CNY180M; $Var[C_1]$ = (CNY42M)^2; revenue cap growth per year, $E[f]$ = CNY27M; $Var[f]$ = (CNY9M)^2; interest rate, $r = 15\%$ per annum; concession period, $n = 25$ years. In Song et al. [12], a single volatility has been assumed; with revenue, revenue growth, revenue cap and revenue cap growth all random variables, comparable standard deviations of 30%, 40%, 25% and 30%, respectively, of their expected values have been assumed here. Results (III) and (IV) of Appendix 6.8 apply. The revenue,R_i, and the revenue cap, C_i, $i = 1, 2, \ldots, 25$, could be anticipated to be independent, and this is the case assumed in the calculations here, but the calculations are essentially the same should that not be the case.

Figure 6.5 (solid curve) shows the option value in any year, calculated for β = 80%. The change in the decline rate of the option value, most noticeably around year 10, occurs because E[PW] and Var[PW] decline at different rates over time. The trend, however, remains downward.

Comparison with the literature. While exact numerical comparison is not possible because of the method used in Song et al. [12], the trend in Fig. 6.5, showing how the yearly option value varies with percentage sharing β, is consistent with the trend given in Song et al. [12]. The yearly option value varies with the level of the revenue cap, $E[C_1]$, as shown in Fig. 6.6. This trend is also consistent with that given in Song et al. [12].

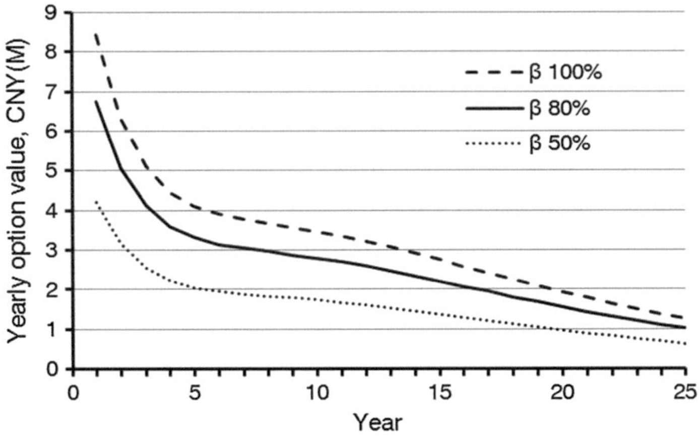

Fig. 6.5 Revenue-sharing example; yearly option value versus time for different β

Fig. 6.6 Revenue-sharing example; yearly option value versus time for different revenue caps; β = 80%

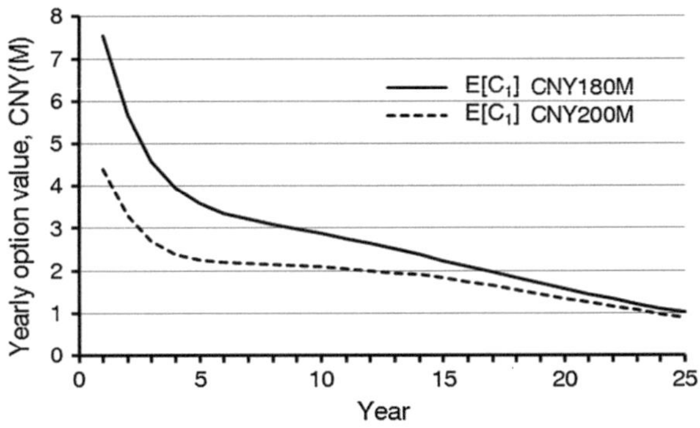

6.4.6 Restrictive Competition Guarantee

Traffic on one road can be influenced by the presence of alternative roads. Alternative roads compete with each other for vehicles and, where roads are tolled, lead to lower revenue. In order to address revenue loss caused by competing roads, a restrictive competition guarantee (non-compete clause) could be used. Here, the authority promises either to not approve any competing road during the concession period, or to provide reimbursement to the concessionaire. This guarantee secures the said road's revenue against loss caused by competing roads.

If the authority approves and/or builds a competing toll road in year i, the concessionaire (option holder) can exercise the option to claim reimbursement from the authority. The authority then compensates the concessionaire a percentage, α, of the revenue shortfall when the revenue is lower than a pre-agreed amount (revenue threshold) in any year. The revenue threshold is adjusted each year. The value of a restrictive competition guarantee is the sum of the option values in each year of the concession period.

In any year $i = T$ (Fig. 6.1), the option is exercised if the revenue shortfall, $X_T > 0$, where: Y_{T1} = the minimum revenue threshold (F_i at $i = T$); and Y_{T2} = the revenue (R_i at $i = T$). The concessionaire receives α percent of the revenue shortfall, that is, αX_T. The option value is calculated based on the present worth derived from an expected value $\alpha E[X_T]$, and a variance $\alpha^2 Var[X_T]$. (This follows from the results in Appendix 6.8.)

Example. Consider an example adapted from Liu et al. [9], using a comparable traffic volume standard deviation equal to 40% of its expected value: revenue in year 1, $E[R_1]$ = CNY207M, $Var[R_1]$ = (CNY76M)^2; revenue growth per year (rate), g_c = 7.5%; minimum threshold at year 1, $E[F_1]$ = CNY140M, $Var[F_1]$ = (CNY20M)^2; threshold growth per year (rate), f_c = 8%; interest rate, r = 5% per annum; concession period, n = 20 years; reimbursement (percentage of revenue shortfall), α = 70%. Results (V) and (VI) of Appendix 6.8 apply. F_i and R_i, $i = 1, 2, \ldots, 20$, are assumed independent, but need not be.

Figure 6.7 (solid curve) shows how the option value varies with time. The upward trend shown is because, in the example calculations, the project revenue increases at a lower rate than the minimum threshold. Accordingly, the option has a higher likelihood of being exercised later in the concession period. Altering the interest rate leads to different option value trends (Fig. 6.7). Higher interest rate values lead to lower present worths, and in turn to lower option values. Figure 6.8 shows how the option value varies with α, the reimbursement percentage of revenue shortfall. Lower α values lead to lower option values, as anticipated. The influence of altering minimum threshold values is shown in Fig. 6.9. The yearly option values increase as the level of the revenue threshold increases.

Comparison with the literature. This book's approach gives a summed yearly option value higher by approximately 0.2% as a proportion of the revenue threshold value,

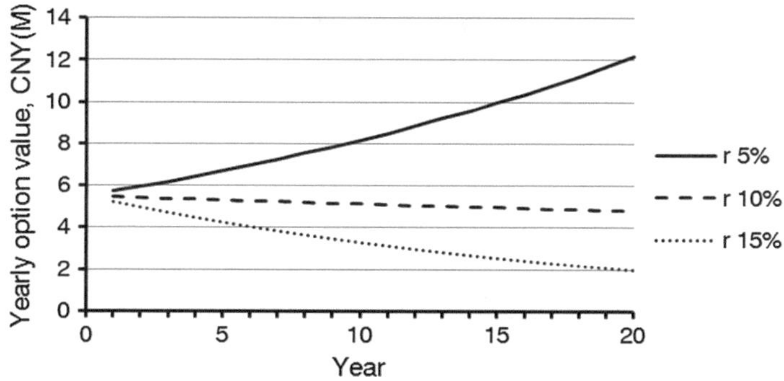

Fig. 6.7 Restrictive competition guarantee example; yearly option value trends with different interest rates

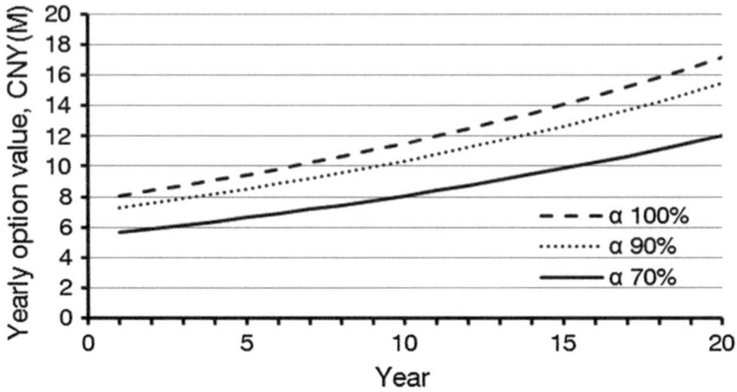

Fig. 6.8 Restrictive competition guarantee example; influence of α

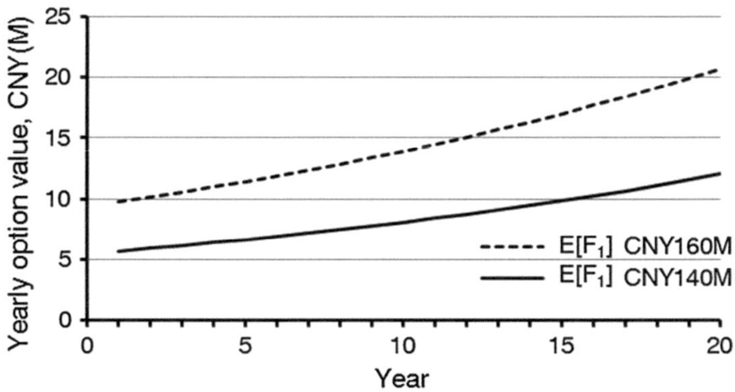

Fig. 6.9 Restrictive competition guarantee example; influence of altering the revenue threshold

when compared with Liu et al. [9]. The summed yearly option value is obtained by using Eq. (1.1) for each year and then adding these values over all years of the concession period. The trend shown in Fig. 6.9 agrees with the observations of Liu et al. [9].

6.4.7 Collar

A collar combines both lower and upper revenue thresholds to create a band. In any year, i, there are two possible options, and depending on the actual revenue, one or neither of these is exercised. The concessionaire holds an option on low revenue. If the actual revenue falls beneath a minimum revenue guarantee or revenue floor, F_i, the concessionaire has a right to claim the revenue shortfall. The authority holds an option on high revenue. If the actual revenue is higher than a maximum revenue guarantee or revenue cap, C_i, the authority has a right to collect the excess revenue. Revenue occurring within the thresholds' envelope or band is unaffected. Alternative to floors and caps in revenue, the situation may be expressed in terms of floors and caps in traffic.

The floor, F_i, and cap, C_i, can be determined in either of two ways, in terms of: zero-cost to both parties; or partial cost to one party, but not to the other party.

No cost to each party can be obtained by setting, in each year, the premium of the concessionaire's option equal to the premium of the authority's option. This is referred to as a zero-cost collar. The two option values are used as proxies for the two premiums. F_i and C_i are adjusted such that the two option values applying to the concessionaire and the authority are the same value. This might be done by first choosing F_i, calculating the associated option value, and then using this option value in a reverse calculation to give C_i.

For the partial cost collar, F_i and C_i are negotiated between the two parties. F_i and C_i can be adjusted to produce a narrower or wider band, with consequent different premiums and option values. Higher F_i and lower C_i separately lead to increased option values. The difference in the premiums represents a cost to one party. However, the parties may agree that there is no up-front cost to either party. In such cases, F_i and C_i might be adjusted according to what might be perceived as a 'fair' allocation of uncertainty to each party.

Using this book's cash flow approach, lower and upper revenue threshold values can be set asymmetrically in each year, and different for all years, without requiring any additional work, assumptions or considerations.

Introduce the superscript notation of c and a for concessionaire and authority, respectively. In any year i = T (Fig. 6.1), the concessionaire's option is exercised if the revenue shortfall, $X_T^c > 0$, where: Y_{T1}^c = the minimum guarantee (threshold) value (F_i at i = T); and Y_{T2}^c = the revenue (R_i at i = T). This is the same as the minimum revenue guarantee above. In any year i = T (Fig. 6.1), the authority's option is exercised if the revenue excess, $X_T^a > 0$, where: Y_{T1}^a = the revenue (R_i at i = T); and Y_{T2}^a = the maximum revenue cap (C_i at i = T). This is the same as revenue-sharing above where β equals 100%.

Example. Consider the zero-cost collar example of Shan et al. [11], using a comparable traffic volume standard deviation equal to 25% of its expected value: traffic in year 1 (vehicles/day × 365 days), $E[v_1] = 25$ k; $Var[v_1] = (6.25\text{ k})^2$; traffic growth per year (rate), γ_c (years 2–5) 6%, (years 6–10) 3.5%, (years 11–35) 2%; toll per vehicle in year 1, $Toll_1 = \$1.30$; toll growth per year (rate), ϕ_c (years 2–5) 5%, (years 6–10) 3%; (years 11–35) 2%; minimum revenue guarantee, year i, $F_i = 78.6\%$ of $E[R_i]$; interest rate, r = 7.5% per annum; concession period, n = 35 years. F_i, here, is deterministic and set as a percentage of the expected revenue (adapted from [6, 7, 11]. (For the minimum revenue guarantee example above, F_i is taken as constant for all i.)

Concessionaire's option. In year 1, for example, $E[Y_{12}^{c}] = -\$11.86M$, $Var[Y_{12}^{c}] = (\$2.97M)^2$, $E[Y_{11}^{c}] = \$9.32M$. Then, $E[X_1^c] = -\$2.54M$ and $Var[X_1^c] = (\$2.97M)^2$ using the results in Appendix 2.11.2. Discounting, $E[PW] = -\$2.36M$, $Var[PW] = (\$2.76M)^2$. From Eq. (1.1), the concessionaire's option value is \$0.296M. The calculation is repeated for other years. Figure 6.10 (solid curve) shows the concessionaire's option value in any year.

Authority's option. The authority's option value, in any year, would ordinarily be calculated based on the expected value, E[PW], and variance, Var[PW], of the present worth of the cash flows associated with the authority's option. However here, the authority's option value is set equal to the concessionaire's option value (just calculated). See Fig. 6.10 for how these option values vary over time. If R_i is assumed to follow a symmetrical probability distribution, then setting F_i and C_i equidistant from $E[R_i]$ will lead to option values for the authority and the concessionaire being the same. Where R_i does not follow a symmetrical probability distribution, then this C_i value will need adjusting.

Comparison with the literature. Comparing available results at years 1, 2, 5 and 10, this book's approach gives option values higher by approximately 1% as a proportion of the concessionaire's minimum revenue guarantee in the corresponding years, when compared with Shan et al. [11]. The yearly option values (same for the concessionaire and the authority) vary with the level of guarantee as shown in Fig. 6.10. The base case referred to in Fig. 6.10 corresponds with the above example values. The alternative case referred to in Fig. 6.10 has the following different values: the expected daily traffic in year 1 is reduced to 23 k vehicles; the traffic growth rate (years 1–5) is reduced to 5%; and the minimum revenue guarantee is increased to 88.3% of the revenue, with the maximum revenue guarantee set equidistant at 111.7%. Moving the guarantee levels closer to $E[R_i]$ increases the option values; while a direct numerical comparison is not possible, this trend is consistent with intuition and the trend given in Shan et al. [11].

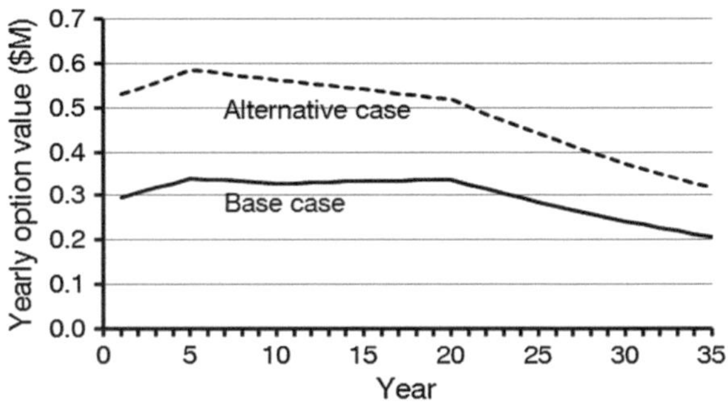

Fig. 6.10 Collar example; yearly option values—concessionaire's (or authority's) options—with different input data. [Alternative case—reduced traffic, reduced traffic growth rate, raised minimum revenue guarantee percentage]

6.4.8 *Traffic Floor and Ceiling*

Traffic floor and ceiling (TFC) is based on pre-agreed lower (floor) and upper (ceiling) traffic levels or thresholds. These two traffic thresholds (floor and ceiling) can be converted to revenue thresholds by multiplying each traffic threshold by the toll ($F_i = TF_i \times Toll_i$, $C_i = TC_i \times Toll_i$). The concessionaire holds the traffic floor option, while the authority holds the traffic ceiling option, as with the collar. Percentages α and β are nominated ($0 \le \alpha, \beta \le 1$), such that if the traffic in any year is lower than the pre-agreed floor level, the floor option is exercised and the concessionaire (option holder) claims a percentage α of the revenue shortfall, while if the traffic is higher than the maximum pre-agreed ceiling level, the ceiling option is exercised and the authority (option holder) receives a percentage β of the revenue excess.

In each year, the TFC involves two options based on lower and upper traffic levels or thresholds (equivalently, revenue levels or thresholds). It is the same as a collar [11], but with two differences—an up-front cost or premium, and a guarantee covering only part of any traffic shortfall or traffic exceedance (a partial coverage guarantee):

- The zero-cost collar has no premium requirement from either party, while the TFC requires premium payments from both the concessionaire and the authority. Premiums, while affecting project viability, do not enter the option value calculations, and hence do not alter the above statements on collar options.
- The collar provides a full coverage guarantee (equivalently, ratios $\alpha, \beta = 1$) above the upper threshold, and below the lower threshold, whereas TFC offers partial revenue protection ($0 \le \alpha, \beta \le 1$).

Negotiation between the parties is needed on the traffic floor and ceiling values, and the lower and upper percentages α and β.

Using this book's cash flow approach, traffic floor and ceiling values can be set asymmetrically in each year, and different for all years, without requiring any additional work, assumptions or considerations.

In any year $i = T$ (Fig. 6.1), the concessionaire's option is exercised if the revenue shortfall, $X_T^c > 0$, where: Y_{T1}^c = the minimum guarantee (threshold) value (F_i at $i = T$); and Y_{T2}^c = the revenue (R_i at $i = T$). The concessionaire receives α percent of the revenue shortfall, that is, αX_T^c. The option value is calculated based on the present worth derived from an expected value $\alpha E[X_T^c]$, and a variance $\alpha^2 Var[X_T^c]$. (This follows from the results in Appendix 6.8.) This is the same as the collar above, but with α introduced. In any year $i = T$ (Fig. 6.1), the authority's option is exercised if the excess revenue above the cap, $X_T^a > 0$, where: Y_{T1}^a = the revenue (R_i at $i = T$); and Y_{T2}^a = the maximum revenue cap (C_i at $i = T$). The authority receives β

percent of the revenue excess, that is, βX_T^a. The option value is calculated based on the present worth derived from an expected value $\beta E[X_T^a]$, and a variance $\beta^2 Var[X_T^a]$. (This follows from the results in Appendix 6.8.) This is the same as the collar above, but with β introduced.

Example. Consider the example of Iyer and Sagheer [6], using a traffic volume standard deviation equal to 25% of its expected value: traffic in year 1 (vehicles/day × 365 days), $E[v_i] = 20{,}654$, $Var[v_i] = (4957)^2$; traffic growth per year (rate), $\gamma_c = 6\%$; toll per vehicle, Toll = INR28.50; traffic floor ratio, Tfr = 80%; traffic ceiling ratio, Tcr = 130%; traffic floor at year i, $TF_i = Tfr \times E[v_i]$; traffic ceiling at year i, $TC_i = Tcr \times E[v_i]$; lower coverage ratio, α = 50% of X_i^c; upper coverage ratio, β = 50% of X_i^a; interest rate, r = 12% per annum; concession period, n = 20 years. Result VII of Appendix 6.8 applies. For example in year 8:

Concessionaire's option. $E[Y|_{82}^c] = -INR323.1M$, $Var[Y|_{82}^c] = (INR78.2M)^2$, $E[Y|_{81}^c] = 80\%$ of $E[Y|_{82}^c] = INR258M$.

Authority's option. $E[Y_{81}^a] = INR323.1M$, $Var[Y_{81}^a] = (INR78.2M)^2$, $E[Y_{82}^a] = 130\%$ of $E[Y_{81}^a] = -INR420M$.

The option values in each year are shown in Fig. 6.11.

Comparison with the literature. This book's approach gives concessionaire's and authority's summed yearly option values lower by approximately 1% and 0.7%, respectively, as a proportion of the sum of revenue threshold values for 20 years, when compared with Iyer and Sagheer [6]. Figures 6.12 and 6.13 show how the concessionaire's and authority's option values vary with Tfr and Tcr, respectively. The yearly option values increase as a result of a higher minimum floor guarantee and a lower maximum ceiling guarantee, and this trend is consistent with the results of Iyer and Sagheer [6].

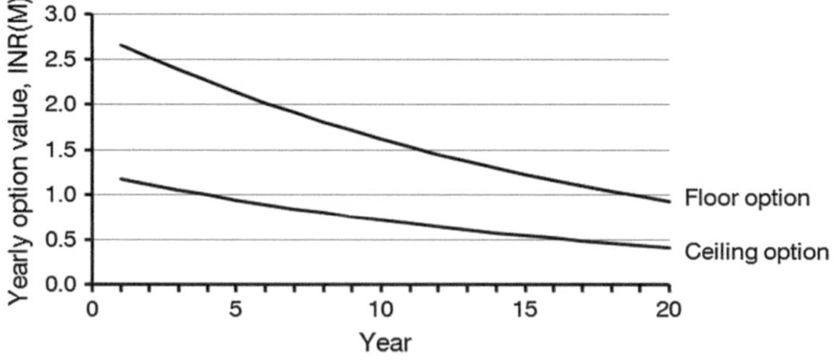

Fig. 6.11 Traffic floor and ceiling example; yearly floor and ceiling option values, Tfr = 80%, Tcr = 130%

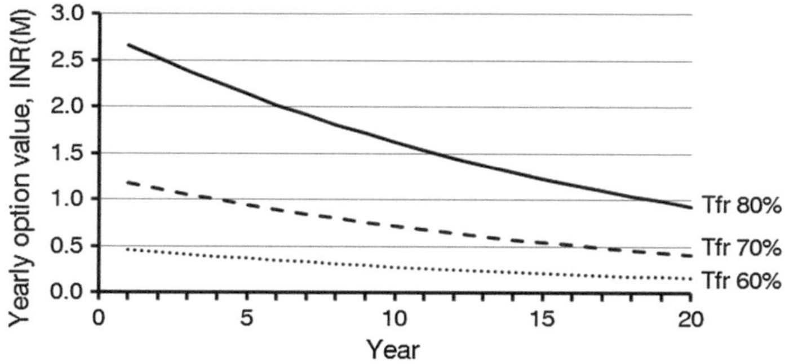

Fig. 6.12 Traffic floor and ceiling example; yearly floor option value; influence of Tfr

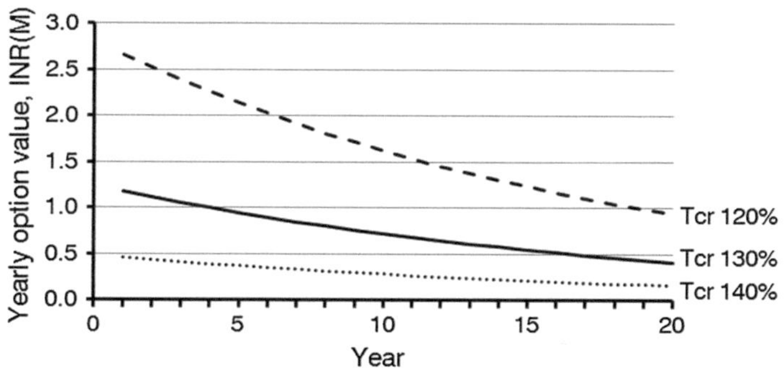

Fig. 6.13 Traffic floor and ceiling example; yearly ceiling option value; influence of Tcr

6.4.9 Toll Adjustment Mechanism

A toll-adjustment mechanism (TAM) is similar in intent to MRG, in that it provides protection for the concessionaire from revenue loss. Whereas MRG gives the concessionaire a guaranteed minimum revenue (the threshold), TAM gives the concessionaire the right to adjust tolls to achieve a revenue level negotiated between the concessionaire and the authority.

Tolls increase annually at a fixed growth rate. However, tolls can be raised further (up to a pre-agreed toll cap) by the concessionaire if the TAM is invoked, that is if the option is exercised. The TAM, if invoked at time $i = T$ when the actual revenue (to the concessionaire) in year T is lower than a pre-agreed level, gives the concessionaire the right to adjust (raise) the toll for the remaining periods $T + 1, T + 2, \ldots, n$. In any year, the toll cap represents the maximum toll that the concessionaire may charge, whereas the pre-agreed minimum revenue level defines the TAM exercise trigger. Both the toll cap and the pre-agreed minimum revenue level are included in the PPP agreement, and are established by negotiation between the concessionaire and the authority. The exercising can only be done once within the concession period—either (depending on the PPP agreement): at a pre-defined year; or in any year. Result VIII of Appendix 6.8 applies. It could be assumed that the traffic would decrease as the toll increases. Accordingly, the traffic with TAM adjustment (raises the toll) might be estimated lower than the traffic without TAM.

In any year i = T (Fig. 6.1), the option is exercised, and the tolls raised for the remainder of the concession period, if the actual revenue in year T, R_T, is less than the agreed minimum revenue threshold. Then: Y_{T1} = the revenue (R_i^{wi}) for years i = T + 1, T + 2, …, n discounted to year T; Y_{T2} = the revenue (R_i^{wo}) for years i = T + 1, T + 2, …, n discounted to year T; and $X_T = Y_{T1} - Y_{T2}$ = revenue difference (with and without).

Example. Consider the TAM example of Chen et al. [5], scenario 5, using a comparable traffic volume standard deviation equal to 20% of its expected value. *Without adjusted tolls:* traffic in year 1 (million vehicles), $E[v_1^{wo}] = 73$; $Var[v_1^{wo}] = 14.6^2$; traffic growth per year (rate), γ_c^{wo} = 2% of v_1^{wo}; toll per vehicle in year 1, $Toll_1^{wo}$ = HKD18; toll growth per year (rate), ϕ_c = 6% of $Toll_1^{wo}$. *With adjusted tolls (up to toll cap—an upper limit):* traffic in year 1 (million vehicles), $E[v_1^{wi}] = 36.5$; $Var[v_1^{wi}] = 7.3^2$; traffic growth per year (rate), γ_c^{wi} = 2% of v_1^{wi}; toll cap per vehicle in year i, $Tollcap_i^{wi}$, (all HKD) 20 (years 2, 3), 25 (4, 5), 30 (6, 7), 35 (8, 8, 10, 11), 45 (12), 55 (13), 65 (14), 75 (15), 85 (16–21), 100 (22–30); interest rate, r = 12% per annum; concession period, n = 30 years. The superscripts wi and wo denote with-invoking and without-invoking TAM, respectively. Result IX of Appendix 6.8 applies.

Assume that the TAM is exercised in year 10, and the toll cap applies in the following years. Then E[PW] = −HKD3,011M, and Var[PW] = $(HKD3022M)^2$. From Eq. (1.1), OV = HKD253.4M. For other years of exercising, Fig. 6.14 shows the corresponding option values. The plot's shape is influenced by the toll cap altering over time.

Comparison with the literature. The maximum yearly option value (year 11) is approximately 2% greater than that given in Chen et al. [5], as a proportion of the corresponding and equivalent minimum revenue threshold. Adapting two more

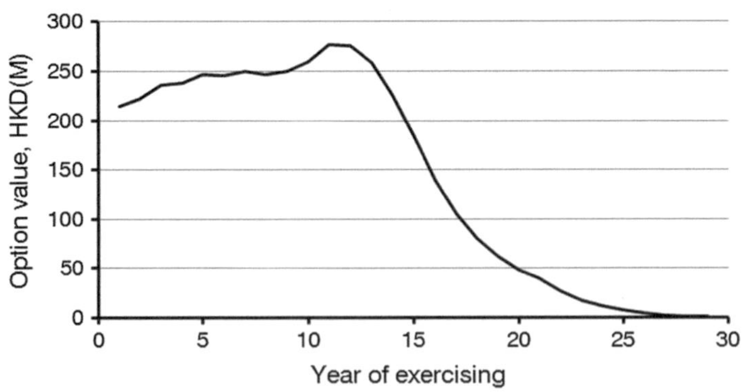

Fig. 6.14 TAM example; option value versus year of exercising

assumed scenarios (scenarios 6 and 8) of Chen et al. [5], the yearly option values increase as the level of traffic volume and toll caps increase, and this trend agrees with the observations in Chen et al. [5].

6.5 Discussion

For all the different option cases analysed above, when compared with existing literature relevant to each case, this book's approach gives option values essentially the same. Supplementary sensitivity-style analyses, conducted by altering the values of the case variables, showed option value trends to be the same with the existing literature. The differences between this chapter's option values and those of existing publications, as a proportion of similar exercise costs, is less than a few percent. Exact agreement would not be anticipated, because of different assumptions applying between those in this book's approach (using variance) and those of existing publications (using volatility). Nevertheless, in this chapter's analysis, assumptions as compatible as possible with the existing literature are made.

A main difference between this book's approach and that of the existing literature is that this book's approach was developed especially for real options and uses conventional discounted cash flow thinking familiar to many. This is compared with the existing literature, which uses methods from the financial markets literature and draws analogies between underlying market variables and infrastructure variables. The book's approach accommodates uncertainty through the use of variances rather than volatility as in the existing literature. Volatility choice is based on analogies with the financial markets methods, and the term may not have direct transference to infrastructure.

The probability distribution for present worth can be assumed to be any appropriate distribution, though a normal distribution was used for convenience in the above calculations. Users of the book's approach are free to adopt an asymmetric probability distribution if they believe that reflects present worth more appropriately. The book's approach can deal with cash inflows and outflows with different levels of correlation, and different over time (that is, varying over the concession period), deterministic cash flows and interest rate variability and uncertainty.

The option valuation is done from the option holder's point of view, whether this is the concessionaire or the authority. Cash flows are established from the option holder's viewpoint. Cash flows, typically, are in terms of those that would exist without exercising the option, and those resulting from exercising the option. The cash flows, depending on the guarantee or agreement, may only be for the year in which the option applies, or over the years extending from the year of exercising the option to the end of the concession period. The approach is the same irrespective of the PPP guarantee or agreement, for example whether a minimum guarantee or a maximum guarantee. There is also no need to distinguish option type, as occurs in the financial markets literature.

6.6 Closure

Guarantees within PPP road project agreements are used by both the authority and the concessionaire to assist in improving project viability and to deal with risk and fairness. The guarantees are one way of addressing the uncertainty in usage or demand that is experienced by roads. Such guarantees can be analysed as options. Heretofore, every different PPP guarantee was presented in terms of its own one-off options analysis, relying on the high level of mathematical skills of the presenters.

With this as background, the chapter demonstrated the following:

- All guarantees (minimum revenue guarantee, buyout, revenue-sharing, restrictive competition guarantee, collar, traffic floor and ceiling guarantee, and toll adjustment mechanism) and their options analyses can be treated in the same way.
- The book's probabilistic present worth cash flow approach provides an original unified approach to PPP toll road options.
- The approach makes no unrealistic assumptions, the level of mathematics necessary is minimal, with knowledge of financial market options analysis techniques not being required.
- Both the concessionaire and the authority are readily able to evaluate, using the book's approach, the impact of any guarantees, providing a basis for the parties to negotiate their PPP agreement.
- The approach permits options that can be exercised yearly or discretely throughout a project's concession period, and these can be treated in a common way.
- The book's approach provides a way forward for analysing all PPP toll road options.

6.7 Extensions

The book's approach is extendable to all infrastructure types, not just roads, and to guarantees which are different to those covered in this chapter, and which may be proposed in the future. Heretofore, the possibilities have been limited because of the restrictive mathematics and assumptions of financial markets methods and establishing analogies with infrastructure.

6.8 Appendix: Some Common Results

For a relationship between any general variables, Z and X_s, and general constants, a_s, of the form, $Z = \sum_{s=1}^{m} a_s X_s$, then,

$$E[Z] = \sum_{s=1}^{m} a_s E[X_s] \qquad Var[Z] = \sum_{s=1}^{m} a_s^2 Var[X_s] + 2 \sum_{s=1}^{m-1} \sum_{t=s+1}^{m} a_s a_t Cov[X_s, X_t]$$

These results are used in the chapter for expressions as follows:

I	$R_i = Toll_i \times v_i$
II	$R_{i+1} = R_1 + (i-1)Toll_i\gamma$
III	$R_{i+1} = R_i + (i-1)g$
IV	$C_{i+1} = C_i + (i-1)f$
V	$R_i = (1+g_c)^{i-1}R_1$
VI	$F_i = (1+f_c)^{i-1}F_1$
VII	$Y_{i2} = (1+\gamma_c)^{i-1}Y_{12} \quad Y_{i1} = (1+\gamma_c)^{i-1}Y_{11}$
VIII	$R_i^{wo} = Toll_i v_i^{wo} \quad R_i^{wi} = Tollcap_i v_i^{wi}$
IX	$R_i^{wo} = Toll_i^{wo} v_i^{wo} = (1+\phi_c)^{i-1}(1+\gamma_c^{wo})^{i-1}Toll_1^{wo}v_1^{wo}$ $R_i^{wi} = Toll_i^{wi} v_i^{wi} = (1+\gamma_c^{wi})^{i-1}Tollcap_1^{wi}v_1^{wi}$

References

1. Brandao LET, Saraiva E (2008) The option value of government guarantees in infrastructure projects. Constr Manag Econ 26(11):1171–1180
2. Carmichael DG (2014) Infrastructure investment: an engineering perspective. CRC Press and Taylor and Francis, London
3. Carmichael DG (2016) A cash flow view of real options. Eng Econ 61(4):265–288
4. Carmichael DG, Nguyen TA, Shen X (2019) Single treatment of PPP road project options. ASCE J Constr Eng Manag 145(2):04018122-1–11
5. Chen Q, Shen G, Xue F, Xia B (2017) Real options model of toll-adjustment mechanism in concession contracts of toll road projects. J Manag Eng ASCE 34(1):04017040, 11 p
6. Iyer KC, Sagheer M (2011) A real options based traffic risk mitigation model for build-operate-transfer highway projects in India. Constr Manag Econ 29(8):771–779
7. Kokkaew N, Chiara N (2013) A modeling government revenue guarantees in privately built transportation projects: a risk-adjusted approach. Transport 28(2):186–192
8. Lewis NA, Eschenbach TG, Hartman JC (2008) Can we capture the value of option volatility? Eng Econ 53(3):230–258
9. Liu J, Yu X, Cheah CYJ (2014) Evaluation of restrictive competition in PPP projects using real option approach. Int J Project Manage 32(3):473–481
10. Power GJ, Burris M, Vadali S, Vedenov D (2016) Valuation of strategic options in public-private partnerships. Transp Res Part A: Policy Pract 90:50–68
11. Shan L, Garvin MJ, Kumar R (2010) Collar options to manage revenue risks in real toll public-private partnership transportation projects. Constr Manag Econ 28(10):1057–1069
12. Song J, Jin L, Dong W (2016) Excess revenue allocation for build-operate-transfer highway projects. J Transp Econ Policy 50:304–324

Chapter 7
Project Delivery—PPP Concession Periods

7.1 Introduction

Public-Private Partnerships (PPPs, P3s) and related concessional delivery methods have current popularity in the delivery of infrastructure across the world, but have been used in different disguises for several centuries. A defining characteristic of a PPP is a concession period, agreed between the two main parties—a public authority and a private concessionaire. The concession period is the time given to the concessionaire to design and construct the infrastructure and collect revenue, before handing back the infrastructure to the authority. This period may be, for example, 20, 30, 40 or 50 years and is different for each project, but is usually long enough at least to fully amortize the initial investment. A concession period that is too long financially benefits the concessionaire in the later years, at the expense of the public. A concession period that is too short does not allow the concessionaire to fully recoup its initial investment or provide a suitable return for the investment.

The issue becomes—what is a reasonable concession period to include in any PPP agreement? For toll roads, an appropriate concession period should allow for the concessionaire to receive a reasonable return for its investment, but not be too long such that the motoring public pays what are perceived as excessive tolls over a perceived excessive period. Determining an appropriate concession period would appear to be crucial in the setting up of PPP delivery of a project. Fixed concession periods, as are currently used, lack the flexibility to deal with this 'long-short' conflict, or to accommodate future uncertainties.

There is anecdotal evidence and public comment, which suggests that concession periods granted on PPP toll road projects are too long in many instances. In effect, the concessionaire receives payback for its investment well within the concession period, while the public continues to pay tolls for longer times than it should. This position excludes those projects where demand estimates, prior to the project, fall short of

D. G. Carmichael, *Future-proofing—Valuing Adaptability, Flexibility, Convertibility and Options*, Management in the Built Environment,
https://doi.org/10.1007/978-981-15-0723-6_7

actual demand once the road is in operation; such estimates may be wrong because of errors in interpreting demand survey data and/or the public's price elasticity of demand behaviour.

There is also the puzzling situation surrounding the use of discounted cash flow (DCF) techniques such as present worth (or net present value) and internal rate of return, which are used to establish the viability of a project. In performing a discounted cash flow analysis, at commercial levels of discount rates, cash flows far into the future have very little present worth. For example, the present worth factors (at 10% per annum) are 0.62, 0.39, 0.09, 0.01 and 0.00 for 5, 10, 25, 50 and 100 years respectively into the future.

And so the question that arises is: why are concession periods longer than say 20–25 years used? Cash flows beyond approximately 20–25 years have small present worth (even allowing for future uncertainty), and contribute little to present day decisions based on any discounted cash flow analysis. Typical toll roads have net cash outflow initially followed by many years of net cash inflow, and so it is the future years of net cash inflow which have small present value, but it is tolls, in these same future years, which are perceived as being unnecessary by motorists.

The chapter suggests a way forward, providing flexibility in establishing the period over which the concessionaire collects revenue. The value analysis to the authority of taking over control of the road is done through an options treatment, and incorporates estimates of uncertainty. Only the financial aspects of PPP agreements are dealt with; options involving some physical aspect of roads are a separate matter—Chap. 4.

The chapter proposes that in a PPP agreement: (i) no explicit fixed concession period be stated, but rather it is flexible and established based on actual project demand and costs; and (ii) beyond a calculated point in time, an option (a right but not an obligation) becomes available to the authority to take over the operation of the road. The chapter's approach eliminates the controversial aspects existing in current PPPs associated with the lengths of concession periods. As a by-product, the approach additionally eliminates another controversial matter, namely that surrounding the setting of tolls and in particular toll formulae and toll adjustments over time, as well as the toll-demand-concession period trade-off. With the chapter's suggested approach, tolls would be perceived as now being necessary by the public rather than excessive.

The chapter's outline is as follows. A literature review covers how concession periods are established, and the optimizing of concession periods. The implications of variability and uncertainty leading to establishing probabilistic payback periods at pre-investment time, as well as the option value of taking over ownership by the authority, are examined. The chapter's suggested methodology involving a flexible concession period follows. A discussion and conclusion follow. A case study example is carried through this chapter.

The chapter is original in the way it addresses variable concession periods, options associated with concession periods and probabilistic concession periods. The chapter will be of interest to anyone dealing with PPPs.

7.2 Background

7.2.1 PPPs

PPPs are said to have a number of potential benefits, most notably the absence of the requirement for public money. But there are disadvantages: the need for management and performance measurements to be put into place to monitor the project; loss of public control and flexibility in making the necessary policy adjustments for future demand and contingencies; the potential for non-compete clauses obstructing future infrastructure projects; and excessive discretion possibly being granted to the concessionaire in regard to setting and regularly increasing the operational road-user charges. Flexible PPP agreements, as proposed here, could partly address these downsides.

In establishing any PPP agreement, uncertainties need to be addressed. These relate to: construction delays and cost overruns; operating and maintenance costs; demand, interest rates, inflation, taxation, competition, sources of financing, insurance costs; and unforeseen events. Many of these are the responsibility of the concessionaire.

7.2.2 Alternative Structures

Over recent years, attempts have been made to address risk to the authority and risk to the concessionaire, primarily looking at revenue allocation resulting from uncertain road usage or demand. However, little attention has been paid to addressing the concession period's influence on a fairer outcome of a PPP to both parties—concessionaire and authority (equivalently the public). Nombela and de Rus [8], Vassallo [13] and Engel [5] write on flexible concession period contracts, in order to avoid renegotiation, but based on auctions and the initial concessionaire bids. None discuss the usage of options as for example discussed in Quiggan [9] or as detailed here. Engel mentions actual PPP projects incorporating variable concession periods in a number of countries, however, nothing appears to have been published on these.

Alternatives to focusing on concession periods, in an attempt to get a PPP agreement that is fair to both parties, have proposed introducing mechanisms such as minimum revenue guarantees and revenue sharing into the PPP agreement. Such mechanisms target patronage or demand, and lower and upper thresholds on revenue and traffic. However, such mechanisms can create contingent issues for the authority; the concessionaire may focus instead on minimizing costs and jeopardize the quality of the infrastructure, rather than attracting patronage. Mechanisms such as minimum revenue guarantees and revenue sharing can also incentivize tenderers to alter their proposed financial models. The tenderer may propose overly optimistic traffic forecasts, or higher discount rates in order to alter thresholds in the PPP agreement, even if the assumptions are inaccurate. As well, they do not address the view expressed here about the public paying perceived excessive or needless tolls towards the end of a concession period.

Buyout options have been discussed in the literature, for example an authority-owned option, with a zero exercise price, to take over the road before the end of the concession period if the concessionaire's after tax internal rate of return (IRR) exceeds a defined percentage; the gain to the authority at time of exercising is the present worth of the remaining cash flows (from the time of exercising till the end of the concession period). This may be referred to as a conditional buyout option, and its value may be low. There has also been described an option to buyout for a predetermined exercise price; exercising only occurs if the present worth of the remaining cash flows exceeds the exercise price. However, it is anticipated that this exceedance would only turn out to be small because it would not be in the concessionaire's interest to negotiate an exercise price that would lead to otherwise.

7.2.3 Concession Period

The concession period can be one of the most important parameters in establishing the financial viability of a PPP project. It is to the advantage of the concessionaire, with current practice, to negotiate for a long concession period and have discretion over setting road-user charges in order to secure loans and service interest costs, while achieving a desirable return. However, if the concession period is too long, the concessionaire is able to gain long-term revenue at little cost, while the authority or the public pay unnecessary long-term road-user charges. Road-user charges, that regularly increase, are disliked by road users. Accordingly, with current practice, the authority should ideally negotiate the shortest concession period possible and that little discretion be given to the concessionaire in setting road-user charges.

Commonly, deterministic present worth (or net present value), internal rate of return and payback period thinking is used by the concessionaire to establish minimum concession periods. To acknowledge uncertainty, sensitivity analysis and Monte Carlo simulation might be used.

A sensitivity analysis demonstrates how altering an input parameter affects the output of a financial model. However, it says nothing of the likelihood attached to this alteration. A sensitivity analysis might be used for example to look at the influence of varying road-user charges and concession period on internal rate of return (IRR). Monte Carlo simulation attempts to capture uncertainty in arriving at a distribution for present worth.

In summary, the literature has undertaken a range of analyses for concession periods. Because of future uncertainties, any concession period established pre-PPP agreement may turn out to be inappropriate based on actual project demand and costs. Publications, generally, have not looked at defining a concession period based on actual project demand, costs and interest rates, and in this respect concession periods which fairly treat both parties—the authority (equivalently, the public) and the concessionaire. This is treated in this chapter.

7.3 Case Study Outline

A case study is developed here to illustrate relevant computations. The project involves a six-lane Vietnamese highway, 105.5 km in length, and involves an investment of approximately 45,500B Vietnam Dong ($2B). The 3-year design and construction period is followed by a 28-year operating period.

The revenue from toll collection alone was considered to be not enough to recoup the initial investment and operating costs. In order to improve the project viability, the authority offered the concessionaire privileges, including:

- Reimbursement of land clearance costs;
- Toll collection from an existing adjacent road; and
- The right to invest and operate service stations along the highway.

The revenue from these sources is additional to the main toll revenue of the highway. Collectively, the revenue from these additional sources is equivalent to a minimum guaranteed revenue.

The concessionaire's costs include operation and maintenance costs, finance/investment costs and corporate tax. An interest/discount rate of 8.4% per annum (p.a.) (based a weighted average value of the interest rates from all funding sources), constant for all years, is adopted in the example calculations below.

Variability estimates in highway toll revenue are established through published traffic volume growth scenarios—low, medium and high. To estimate toll revenue variances (and expected values), the three scenarios are regarded as pessimistic, most likely and optimistic values respectively, as is outlined in Chap. 2. Variances in the revenue from the additional sources are obtained by assuming similar estimate ranges as that assumed for highway traffic variability, namely ±50% either side of most likely values. The traffic revenue and additional revenue are assumed perfectly correlated in order to establish the variance of the total revenue coming to the concessionaire. Variability in costs was estimated from available concessionaire information.

7.4 Notation

Specific notation used in this chapter is as follows:

I	year counter, $i = 0, 1, 2, \ldots, t$
M	mean of the $PW\|_t$ distribution to the right of R measured from $PW\|_t = R$
$PW\|_t$	collective (summation) present worth over years $0, 1, 2, \ldots, t$
R	agreed return to the concessionaire for the investment.

(Note that R is not the same as the usual return on investment, ROI.)

Y_{i1} revenue in year i

Y_{i2} costs in year i

Φ $P[PW|_t > R]$

For pre-investment estimates, X_i, Y_{i1}, Y_{i2} and $PW|_t$ are random variables.

7.5 Pre-investment Analysis

7.5.1 Outline

In broad terms (made more definite below), the chapter is proposing: (i) that no concession period be stated up front (pre-investment); (ii) the concessionaire operates the road until an agreed return, denoted R, is achieved; and (iii) there be available an option (a right but not an obligation) for the authority to take ownership of the road after this return is achieved. (Note that R is not the same as the usual ROI, return on investment.)

The background and ramifications of this in terms of project management can be demonstrated through a pre-investment (up-front) analysis. Such an analysis can be used to establish up-front estimates (non-binding between the parties) of: project viability and an approximate concession period that could be anticipated; the value to the authority (or road user) of holding an option to take over ownership of the road; and the cost to the road user of longer concession periods.

For the up-front analysis, in order to capture the future uncertainty, a probabilistic analysis is used, allowing for the revenue and cost estimates in each year to contain this uncertainty. That is, revenue and costs are probabilistic.

In a similar fashion to Appendices 2.11.1 and 2.11.2, let X_i be the net cash flow in any year i, $i = 0, 1, 2, \ldots, n$. It is the result of two cash flow components, the total revenue Y_{i1} and the total cost Y_{i2}, $X_i = Y_{i1} - Y_{i2}$. For uncertainty in the cash flows,

$$E[X_i] = E[Y_{i1}] - E[Y_{i2}]$$

$$Var[X_i] = Var[Y_{i1}] + Var[Y_{i2}] - 2\rho_{12}\sqrt{Var[Y_{i1}]}\sqrt{Var[Y_{i2}]}$$

where ρ_{12} is the correlation coefficient between Y_{i1} and Y_{i2}.

Let $PW|_t$ be the discounted cumulative sum (present worth) of the actual yearly net cash flows (revenue minus costs) up to and including year t, $PW|_t = \sum_{i=0}^{t} \frac{X_i}{(1+r_i)^i}$. Then,

$$E[PW|_t] = \sum_{i=0}^{t} \frac{X_i}{(1 + r_i)^i}$$

$$\mathrm{Var}[\mathrm{PW}|_t] = \sum_{i=0}^{t} \frac{\mathrm{Var}[X_i]}{(1+r_i)^{2i}} + 2\sum_{i=0}^{t-1}\sum_{j=i+1}^{t} \frac{\rho_{ij}^{x}\sqrt{\mathrm{Var}[X_i]}\sqrt{\mathrm{Var}[X_j]}}{(1+r_i)^{i+j}}$$

where ρ_{ij}^{x} is the correlation coefficient between X_i and X_j. In the following case example calculations, on the basis that the net cash flow in one year will be determined by the same influences as in the next year, high correlation could be assumed for years close to each other and decreasing slightly as years get further apart. In the following, ρ_{12} and ρ_{ij}^{x} are set as 1, but other values could be used. The above expressions can be extended to include probabilistic interest rates as in Appendix 2.11.3. $\mathrm{PW}|_t$ could be anticipated to follow a normal distribution for larger t, because of the additive cash flow components.

7.5.2 *Approximate Concession Period*

An approximate up-front estimate for the concession period can be established through first looking at the payback period (PBP), albeit a random variable.

$\mathrm{PW}|_t$ can be calculated for $t = 0, 1, 2, \ldots$ years based on estimated cash flows. For usual road cash flows, as t gets bigger the mean of the probability distribution for $\mathrm{PW}|_t$ will become more positive, negative $\mathrm{PW}|_t$ values will become less, and the area under the $\mathrm{PW}|_t$ distribution to the left of the origin will become less. That is, as anticipated, project viability increases with t. For any time t, there will always be negative values of $\mathrm{PW}|_t$, and there will always be an area under the $\mathrm{PW}|_t$ distribution to the left of the origin. Figure 7.1 shows how $\mathrm{PW}|_t$, and the probability that $\mathrm{PW}|_t$ is positive, vary as

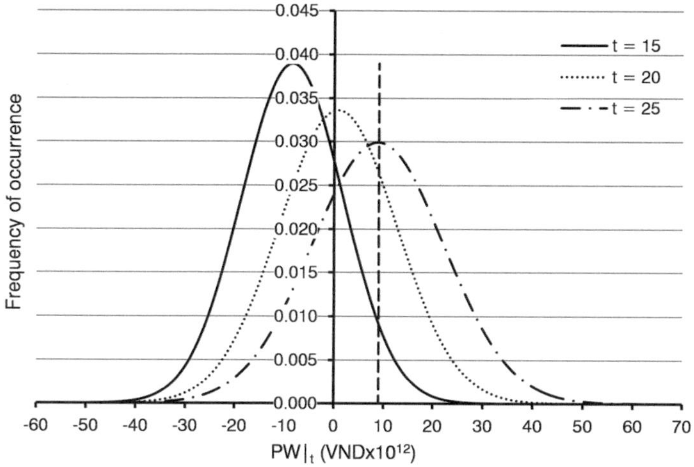

Fig. 7.1 Probability distribution of $\mathrm{PW}|_t$ for different t (years) for case example; R at an example of 20% of total investment shown with dashed line

t varies for the case example; the expected value of and probability distribution for $PW|_t$ move to the right with increasing t, while the variance of $PW|_t$ increases with t (as more cash flows are incorporated into the calculation of $PW|_t$), giving flatter probability distributions. Superimposing on Fig. 7.1 a dashed line corresponding to the value at which $E[PW|_t] = R$ shows the associated probability that $PW|_t > R$ for each value of t.

In the usual deterministic case, the payback period (PBP) is defined as the time t where $PW|_t = 0$, that is PBP = t when $PW|_t = 0$. It follows for the probabilistic case (pre-investment analysis) that the probability that PBP is less than an assumed t is the same as the probability that $PW|_t$ is negative. $P[PBP > \text{nominated } t] = P[PW|_t < 0 \mid \text{nominated } t]$. Then the cumulative distribution function for PBP is obtained from $1–P[PBP > t]$.

For practical computation, this approach would appear satisfactory. In terms of related calculations of internal rate of return, *it is good enough for most practical purposes* [7]; some special circumstances are noted by Hillier [6].

The distribution for PBP may be found numerically. For each of a series of values of t, $E[PW|_t]$ and $Var[PW|_t]$ are obtained leading to a probability distribution for $PW|_t$; from each distribution, a value for the cumulative distribution for PBP is obtained as above, and subsequently the probability density function for PBP is obtained either by differentiation of the cumulative distribution function, or by assuming PBP follows a normal distribution. It is argued by Hillier [6] that if the probability distribution for $PW|_t$ is normal, then so the probability distribution for internal rate of return will approximate that of a normal distribution. This argument should extend to payback period.

Figure 7.2 shows the cumulative distribution function for the payback period corresponding to the case example. The concession period will be R shifted from this.

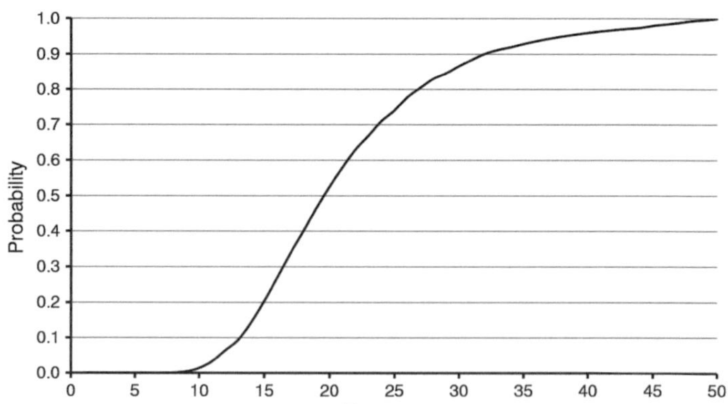

Fig. 7.2 Cumulative distribution function for payback period; case example

7.5.3 *Option Value*

The chapter proposes that the authority hold an option (a right but not an obligation) to take over the toll road at some time in the future, but it may not necessarily exercise that option. If the authority's option is exercised at the time at which $PW|_t = R$, then the concessionaire only operates and receives revenue from the road from $i = 0$ to this time. After this time, the authority is able to own and operate the road. Should the option not be exercised by the authority, the concessionaire continues ownership and operation.

In Fig. 7.1 for the case example, probability distributions of $PW|_t$ for different t are shown, while the vertical dashed line is located at the value where $E[PW|_t] = R$. Calculating the loss to the road user by the authority not exercising its option, is equivalent to calculating the option value based on the tails of the above distributions to the right of the dashed line. Based on Eq. (1.1), this option value is $OV = \Phi M$, where Φ is $P[PW|_t > R]$, P is probability, and M is the mean of the $PW|_t$ distribution to the right of R measured from $PW|_t = R$.

The expected value and variance of, and distribution for $PW|_t$ vary with t, as exampled in Fig. 7.1. The vertical dashed line (value of R) in Fig. 7.1 may also vary with t depending on the PPP agreement. The loss to the road user increases with time if the authority's option is not exercised, because this option value increases with time. The longer it takes for the authority to exercise the option, then the greater is the cost to the road user. This is also intuitive, but the chapter's method establishes the real quantitative value of the cost to the road user. The loss to the road user differs over time as in Fig. 7.3.

Road-user losses to the right of the dashed line in Fig. 7.3 are avoidable through the authority exercising its option. Later exercising (equivalently, longer concession periods) leads to higher road-user losses, and higher gains to the concessionaire.

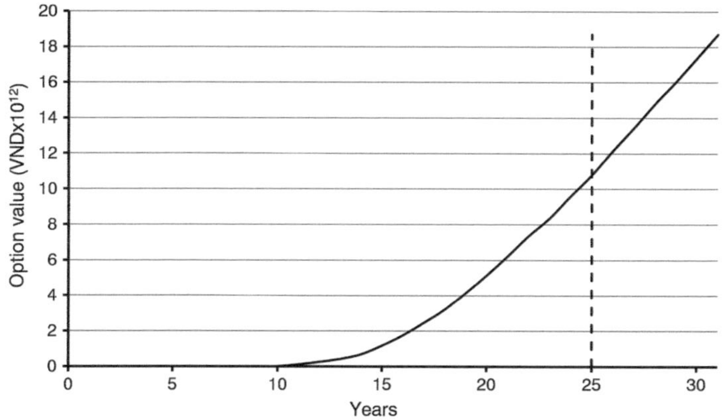

Fig. 7.3 Loss to the road user over time with non-exercising of the authority's option; R-determined year shown with dashed line—case example

7.6 Proposal—Flexible Concession Period

7.6.1 Outline

In existing practice, concession periods could be anticipated to be negotiated as part of risk management for both the authority and the concessionaire. If generous contingencies are included to take care of uncertainties, concession periods could be anticipated to be adopted greater than what would be desired by the road user.

In this chapter, it is proposed that the concession period be flexible, that is, it is not fixed up front (pre-investment), but rather is determined with time, based on actual project costs and revenues. In particular, the concession period is determined such that the concessionaire has a guaranteed and fair way of getting a return for its investment, yet does not burden the road user long onto the future.

There is an analogy here with a lump sum contract (analogous fixed concession period) and a cost reimbursable contract (analogous flexible concession period). A lump sum price includes a contingency but the price is fixed; cost reimbursement does not include a contingency, but the price is only determined after the work is done. Similar issues involving lump sum payment versus cost reimbursable payment will transfer across this analogy to fixed concession periods versus flexible concession periods, and disclosure and verification of costs in the cost reimbursable case.

This chapter is proposing that the concession period be calculated relative to an actual (not pre-calculated) payback period. The payback period represents a recovering of costs only and does not give a concessionaire a return above its investment. This is so unless the interest rate used incorporates a margin for this return to give a discount rate. (Carmichael [2], among others, notes the irrationality of using discount rates in cases involving uncertainty, and so this approach is not adopted here. Rather, this chapter suggests using a basic interest rate, equivalent to what some publications might call a 'risk-free' rate.)

No prospective concessionaire will invest where the concession period is less than the payback period, but rather a concessionaire wants something extra—a positive 'return', here denoted R. (Note that R is not the same as the usual ROI, return on investment.) Section 7.5 (pre-investment analysis) uses probabilistic estimates to determine an estimate of the period over which the concessionaire operates the road. To establish the actual concession period, the calculation of $PW|_t$ is done yearly into the future based on actual cash flows. $PW|_t$ is now deterministic, compared to the up-front analysis where it is probabilistic. Then, the chapter's proposal is:

No concession period is stated up front (pre-investment). The concessionaire operates the road provided $PW|_t < R$, *where* R *is a negotiated and nominated concessionaire's return for the investment. For* $PW|_t \geq R$, *the authority can (but does not have to) exercise an option to take over ownership. The option cannot be exercised for* $PW|_t < R$.

The concessionaire operates the road for as long as it takes to get the nominated return, R. The length of time it takes for the concessionaire to get this nominated return becomes equivalent to conventional thinking of what a concession period is.

An estimate of this period of time is possible pre-investment (Sect. 7.5), and this estimate becomes more refined over time when the road is in operation and actual cash flows become known.

7.6.2 *Operational Matters*

The above proposal creates the following operational discussion points.

1. The formula for calculating the return, R, is selected by agreement of the concessionaire and the authority up front (pre-investment). Some possibilities are:

- As a percentage of initial investment cost;
- As a fixed amount, possibly allowing for some escalation based on inflation and some consumer price rise/fall index; or
- As variable with time.

Clearly, the greater the value of R selected, the longer the concession period. A pre-investment analysis (Sect. 7.5), based on cash flow estimates, can be used to give a guideline approximate concession period that could be anticipated.

A variable percentage could be defined in a number of ways. For example, let R increase by p% per year, a pre-defined and agreed amount. Then $R_{t+1} = R_t(1+p\%)$.

Shen et al. [10], Shen and Wu [11], Shen et al. [12] suggest that the additional return to the concessionaire (calculated pre-investment) be the product of the concessionaire's capital investment and the concessionaire's desired rate of return. Desired rates of return depend on the concessionaire's opportunity costs, risks and risk attitude, among other factors.

2. The interest rate, r_i, used in each year i for discounting purposes is selected by agreement of the concessionaire and the authority for each year, and selected in that year. This interest rate contains no adjustment for the concessionaire's business 'risks' or adjustment for the road project 'risks'.

3. Open book accounting is necessary for both revenue and costs, such that the actual net cash flow (revenue minus costs) in any year i, X_i, can be agreed each year between the concessionaire and the authority. The X_i are only privy to the concessionaire and the authority, and are not publicly available. This requires no new thinking over what is currently required for PPP guarantees such as minimum revenue or revenue sharing (Chap. 6).

4. With X_i and r_i agreed as each year i passes, it is possible to calculate $PW|_t$ each year, and hence establish, on an on-going basis, whether $PW|_t$ is less than, equal to or

greater than R. With each year of operation, an updated estimate of the approximate concession period becomes possible.

5. The concessionaire and the authority need to agree on the accounting principles used. The chapter's suggested approach raises issues that are similar to those that occur in cost reimbursable contracts. For such contracts there is a need for agreement between the parties on what is reimbursable and what is not reimbursable, and how overheads and profit are handled. Here there is a need for agreement as to what are costs and what are revenues. There is also a need to think of the concessionaire's business profits and business overheads and how they might be incorporated into R, such that only project overheads get included in the project costs.

6. Having a concession period undetermined at the project start may raise issues about the level of quality that the concessionaire builds into the road. Typically, the road will have to satisfy certain performance standards on handover, at the expiration of the concession period, to the authority. The road design, construction and maintenance practices will therefore be based on this time period, which is now variable. However, an approximate concession period can be established up front (pre-investment) provided sensible estimates are used for future costs and revenue (Sect. 7.5).

7. Used by itself, the approach suggested is more suitable for high-demand roads. For low- or variable-demand roads, the approach can be used in conjunction with any guarantee that the parties may wish to use, for example, a minimum revenue guarantee (Chap. 6). Such guarantees could be structured to give a reasonable payback period, and not have the concessionaire operate the road indefinitely.

8. Care will need to be exercised by the authority overseeing the road design and construction, and the period allowed for these activities. There will need to be an agreed scope and quality statement, and an agreed time period for design and construction. Without these agreed up front (pre-investment), there is the potential for the concessionaire to manipulate the outcome.

9. There is also potential for manipulating the operating and maintenance levels, which in turn influences cash flows. There will need to be agreed operation and maintenance parameters. For example, closing lanes leads to lower road-user comfort, and possibly preference for travelling on alternative roads.

10. It is anticipated that for usual road cash flows, the present worth of all cash flows is negative in the initial years, and becomes positive in later years. Should the cash flows be such that $PW|_t < R$ always, then the ownership and operation of the road remains with the concessionaire. This attempts to address any lender anxiety about not being repaid loaned money. Should this be unacceptable to the authority, then a fall-back (upper limit) concession period could be stated in the PPP agreement,

in which case, the concessionaire would cease operating the road at the end of this fall-back period.

11. It would also be possible for the concessionaire to have an option to abandon the road and transfer its operation to the authority. However, this might only be contemplated if the road was losing money. This situation is not considered here, but could be readily incorporated. The concessionaire always has the ability to sell its interest in the road, even perhaps at a loss.

7.6.3 Case Example

For the case example project, with the road newly completed and actual cash flows and interest rates yet to be recorded, let actual revenues and costs equal their estimate means, and let the interest rate stay constant over time. With R at 20% of total investment, then the concession period becomes approximately 25 years.

7.7 Discussion

The chapter's approach of not stating the concession period up front (pre-investment) avoids controversial aspects of concession periods. The approach avoids concession periods that are too short (disadvantageous to the concessionaire) and avoids periods that are too long (disadvantageous to the authority and the travelling public). The concessionaire's involvement lasts only as long as it takes to reach a fair return for its investment.

The approach also eliminates the controversial matters surrounding the setting of tolls—what are suitable tolls and what are suitable periodic toll adjustments within the PPP agreement, as well as the toll-demand-concession period trade-off. Using the approach, tolls would be perceived by the public as now being necessary rather than excessive. With the chapter's approach, larger tolls translate to earlier payback, earlier ownership of the road by the authority and the public paying over less years.

Forecast traffic demand possibly contains the highest level of uncertainty in a PPP road project and leads to the highest risk. Accordingly, with existing practice, any prescribed up-front concession period will necessarily reflect this high risk. As well, traffic forecast levels may be manipulated to suit whichever party is paying for the forecasts—low levels for the concessionaire, high levels for the authority. The chapter's approach, on the other hand, establishes the equivalent of the concession period but based on actual, not uncertain or biased traffic/demand; the chapter's approach removes the risk and bias from the concession period determination.

Black [1], analyzing eight Australian toll roads and tunnels, showed the average traffic to be 48% of what was originally forecast. This contributes to a general opinion that traffic forecasts are also characterized by optimism bias. Patronage models may be flawed. And, there is the suggestion that the preferred concessionaire tenderer will be the one that has the highest or most aggressive forecasted revenue/traffic, giving potential concessionaires an incentive to manipulate their models. The chapter's approach avoids optimism bias and any other bias.

In many cases in the past, risk associated with uncertain traffic demand was carried by the concessionaire. This risk might be reduced by the inclusion in the PPP agreement of non-compete clauses for juxtaposed arterial roads, as well as decreasing the capacity of alternative routes in the corridor that the PPP road is serving. The chapter's approach avoids such risk deliberations.

The chapter's approach requires open book accounting, trust, cooperation and goodwill between the two parties—concessionaire and authority. Open book accounting is not new in infrastructure delivery, while trust, cooperation and goodwill concepts exist in partnering, relationship contracting, integrated project delivery and alliance ideas, which have current popularity.

It is anticipated that yearly negotiation between the parties will be required to establish agreed actual cash flows and interest rates. In any investment analysis, there can be different viewpoints on what is the appropriate interest rate to use. The chapter's analysis allows the interest rate to vary yearly and, for pre-investment analysis, to incorporate uncertainty (Carmichael and Bustamante [3]; Carmichael and Handford [4]).

7.8 Conclusion

The chapter gave a means by which the time over which the concessionaire controls a PPP road can be established that is fair to both parties. It avoids stating a definite concession period up front, but rather the equivalent of a concession period is established based on actual cash flows as the project progresses. It avoids the need for the concessionaire to include conservatism and contingencies in its feasibility studies because of future uncertainties. (This is the same argument used in contract payment types—fixed price contracts are costed higher than schedule of rates/unit price or cost reimbursable contracts because of the need to include contingencies.)

The chapter's approach eliminates the controversial aspects existing in current PPPs associated with the lengths of concession periods, toll setting and toll adjustments.

The chapter demonstrated how the option value to the authority of taking over control of the road could be calculated. The approach, based on a second order moment analysis, and acknowledging uncertainty, requires minimal mathematical knowledge and is an extension of conventional feasibility analysis.

A case study on a Vietnamese PPP toll road demonstrated the approach and typical values.

7.9 Extensions

The analysis of existing PPP toll road demands and costs is difficult to undertake because public information on toll roads is not generally forthcoming from either concessionaires or authorities. Even within annual reports of companies, information is disguised, and often mixed with taxation, depreciation amortization matters, and 'creative accounting'. More case studies, using divulged data would assist the take-up of the chapter's proposal.

The chapter does not consider PPPs where the public authority collects the tolls (if applicable) and then reimburses the concessionaire. However, the chapter's treatment could be readily altered to suit this case, or to suit the situation discussed by Quiggan [9] where both put and call options exist allowing either party to terminate the PPP agreement.

References

1. Black J (2014) Traffic risk in the Australian toll road sector. Public Infrastruct Bull 1(9):18–29
2. Carmichael DG (2017) Adjustments within discount rates to cater for uncertainty-guidelines. Eng Econ 62(4):322–335
3. Carmichael DG, Bustamante BL (2014) Interest rate uncertainty and investment value-a second order moment approach. Int J Eng Manag Econ 4(2), 176–189
4. Carmichael DG, Handford LB (2015) A note on equivalent fixed-rate and variable-rate loans; borrower's perspective. Eng Econ 60(2), 155–162
5. Engel E (2008) Public-private Partnerships: when and how, MIT presentation, MIT, Massachusetts, pp 39
6. Hillier FS (1965) Supplement to 'The derivation of probabilistic information for the evaluation of risky investments'. Manag Sci 11(3):485–487
7. Hodges SD, Moore PG (1968) The consideration of risk in project selection. J Inst Actuar 94:355–378
8. Nombela G, de Rus G (2004) Flexible-term Contracts for Road Franchising. Transp Res Part A 38:163–179
9. Quiggin J (2005) Public-private partnerships: options for improved risk allocation. Aust Econ Rev 38(4):445–450
10. Shen LY, Li H, Li QM (2002) Alternative concession model for build operate transfer contract projects. ASCE J Constr Eng Manag 128(4):326–330
11. Shen LY, Wu YZ (2005) Risk concession model for build/operate/transfer contract projects. ASCE J Constr Eng Manag 131(2):211–220
12. Shen LY, Bao HJ, Wu YZ, Lu WS (2007) Using bargaining-game theory for negotiating concession period for BOT-type contract. ASCE J Constr Eng Manag 133(5):385–392
13. Vassallo JM (2006) Traffic risk mitigation in highway concession projects: the experience of Chile. J Trans Econ Policy 40(3):359–381

Chapter 8
Project Delivery—CDM Projects

8.1 Introduction

The Clean Development Mechanism (CDM) is one of three flexible emission reduction mechanisms flowing from the 1997 Kyoto Protocol [11], where certified emission reduction credits or units (commonly referred to as CERs or carbon credits) are generated in developing countries for use by industrialised countries (referred to as Annex 1 parties—[11]) in achieving their emission reduction targets. Although the Protocol expired on 31 December 2012, the eighteenth session of the Conference of the Parties (COP 18) in Qatar [12] extended the duration of commitments to 31 December 2020. The aim of CDM projects is to encourage sustainable development and to reduce global emissions at least cost [8, 11]. CERs are generated by the projects for every tonne reduction in carbon dioxide equivalent (CO_2-e) below a defined baseline; this ensures the environmental worth of a CDM project [7, 9]). CERs can be traded and sold in carbon markets, or used by organizations to comply with emission targets (UNEP [8, 9]).

In order to access CERs, a project requires upfront registration, validation and ongoing compliance checking, all of which have transaction costs, additional to regular project costs, detracting from investing in CDM projects. These transaction costs affect project viability. The chapter proposes one way whereby the influence of these transaction costs can be reduced.

The value of CERs fluctuates with the price of carbon in the carbon markets, and CER output generated by a project may fluctuate, and so the project income from CERs contains uncertainty. This, in turn, affects the project's attractiveness. Potential downward price fluctuations might be mitigated with futures contracts in which a CER price at a given date is locked in, however this loses any favourable market price movements. Options can also be held on carbon prices, but not on project CER output. It would appear reasonable therefore to build in the potential for options in CDM projects, whereby the project can benefit from CERs when the CER price

D. G. Carmichael, *Future-proofing—Valuing Adaptability, Flexibility, Convertibility and Options*, Management in the Built Environment,
https://doi.org/10.1007/978-981-15-0723-6_8

and/or project CER output are favourable (upside), but not be put to the expense of ongoing compliance costs when the CER price and/or project CER output are unfavourable (downside). This decreases the financial risk of a CDM project, and makes CDM projects more attractive financially.

In this chapter, the option refers to the choice of whether to claim carbon credits or not depending on fluctuations in the carbon price and/or project CER output. Allowing options gives the investor freedom to walk away from, or take advantage of, the carbon-based revenue side of the project at any given time. Options allow the investor to make choices throughout the project lifespan, and having these choices or flexibility leads to a higher investment value. Options turn the exposure associated with the uncertainty in future carbon prices and/or project CER output to the investor's advantage.

Introducing options would require a CDM rule change. (For an explanation of current rules, refer [1].) Allowing options, in turn, should lead to higher CDM viability and more CDM projects, leading to more global emission reductions.

The chapter's argument is developed through firstly giving the proposal. Some background information on CDM projects and carbon markets follows. The presence of uncertainty in the CER price and project CER output, together with the introduction of options, implies a probabilistic analysis. This book does this through a straightforward second order moment analysis using expected values and variances only of project cash flows [3]. Two registered CDM projects are used to support the chapter's proposal, one on hydropower and one on wind power.

The chapter will be of interest to investors, project developers and policy makers involved with CDM projects, and generally to those concerned about greenhouse gas emission reductions and sustainability. The chapter's idea of allowing options within CDM project rules is original. Policy issues relate to the UNFCCC rules overseeing CDM projects [11].

Only financial matters relating to CER price and project CER output are addressed in this chapter. Technological, political, environmental and administrative and procedural matters, as well as project performance are not discussed.

8.2 Methods

8.2.1 Inclusion of Options

Given the carbon-emission reduction and sustainability benefits of CDM projects, it would appear reasonable that CDM projects should be encouraged. One way to do this is to make CDM projects more financially attractive.

This chapter is proposing that the full extent of options be allowed with CDM projects in order to increase the attractiveness and number of CDM projects. This will involve a CDM rule change. Discussion is given below on any detriment that may flow to any stakeholder from such a rule change.

In the context of this chapter's proposal, namely options in conjunction with CDM projects:

- The option right is to claiming carbon credits (CERs) during the CER crediting period—credits may or may not be (optionally) claimed, depending on the current CER price and/or project CER output.
- The credits obtained represent a positive cash flow; the cost of making a claim for the credits represents a negative cash flow.
- The premium paid, at project outset, is the cost of establishing the project as a CDM project.
- The time of exercising the option is during the CER crediting period; typically each year during this period, a decision is made to exercise the option or not.

Carbon credits will only be claimed (that is, the option will only be exercised) if it is favourable to do so. In years where the price of carbon is high (low) and/or the project CER output is high (low), whereby the return from CERs is greater (less) than the cost of their claiming, the option is (not) exercised.

This flexibility in decision making has a value, and (based on Eq. 1.1) it is always greater than the value of having to always claim credits every year during the crediting period (namely the current CDM practice). That is, allowing options increases the attractiveness of CDM projects.

8.2.2 *Background—Clean Development Mechanism*

8.2.2.1 General

The Clean Development Mechanism, under the Kyoto Protocol of 1997, is an attempt to reduce global carbon emissions at low cost, and promote sustainable development. Developed countries can invest in projects located in developing countries, and receive CERs. However, because of the volatility in the price of carbon, among other things, CDM projects have not reached their full potential, and investors are naturally cautious.

Proposed CDM projects and the issuance of CERs are subject to approval by an Executive Board (EB), in order to ensure that emission reductions are real and additional [1]. Emission reductions are relative to a baseline, which follows from using approved baseline methodologies [11]. Emissions are checked to see that they are below that which would have occurred in the absence of the CDM project (the business-as-usual case), while financial additionality requires demonstrating that the project is not viable without CDM classification (and hence CER revenue). Typically, financial additionality using internal rate of return is checked deterministically, ignoring any uncertainty in carbon price and project CER output. However, when uncertainty is acknowledged, this practice can introduce anomalies [5].

The progress of the Clean Development Mechanism has been impressive in terms of registered projects, and CERs supplied [11], though low prices for carbon are restrictive. Along the way there have been teething issues, and suggestions for streamlining approvals and baseline methodologies have been made in order to reduce transaction costs and speed up project registration. Projects are subject to rigid crediting periods and practices, and are unable to respond to fluctuating carbon prices and/or project CER outputs.

8.2.2.2 Options and CDM Projects

Options in conjunction with CDM projects have been suggested in other publications. However their usage is in a different sense to that in this chapter. Suggestions have been made to: purchasing and selling options in carbon tradable offsets; and developing a negotiation stance, based on risk sharing, between the host country and the investor, where CDM projects are viewed as emissions trading between developed and developing countries. Also, the literature mentions the derivatives market in emissions trading, futures contracts, insurance policies and hedging practices that allow an investor to reduce risk associated with carbon prices. There is nothing published that deals with both fluctuating CER prices and fluctuating project CER output, options and rule changes in CDM projects.

8.2.2.3 Costs

Costs and revenue in the following refer only to carbon-based costs and revenue, and not to total project costs and revenue. Only carbon-based costs and revenue are relevant in the following option calculations.

Transaction costs are (i) upfront or pre-operational, and (ii) ongoing or operational:

- Upfront or pre-operational costs include search costs, negotiation costs, validation costs and approval costs, made up of CDM project cycle costs for preparation and review, baseline study, environmental assessment, stakeholder consultation, approval, validation, consultation and project appraisal, and legal and contractual arrangements. Upfront approval and registration requires the preparation, by the project proponent, of a Project Design Document (PDD), which describes the project and includes information on baseline methodology, emission reduction calculations, duration and crediting period of the project, additionality and sustainability criteria, and environmental impact. The project is put forward to the host country for approval, and to ascertain that the project contributes to the country's sustainable development goals [9]. A Designated Operational Entity (DOE) reviews the PDD. Having been validated, the project can then be registered (up to the time of the project being operational) with the EB for a fee dependent on the annual amount of CO_2-e that is given as being reduced. The fee is paid upfront, but is later deducted from the share of proceeds at the issuance of the CERs.

- Ongoing or operational costs include monitoring costs, certification costs and enforcement costs made up of CDM project cycle costs for sales of CERs, adaptation levy, risk mitigation, verification and EB administration. A monitoring report gives estimates of the CERs generated and is submitted for verification to a DOE, which checks against an approved methodology, and subsequently for approval by the EB. Certification by the DOE, that the project has achieved the emission reductions, leads to the issuance of a verified amount of CERs. The frequency of issuance varies among projects, however once a year is common. There is also an adaptation fee (exempt for least developed countries), calculated at 2% of the CERs claimed; this goes to a fund to help developing countries' climate adaptation.

Of particular relevance to this chapter are the monitoring, verification and certification costs. In monitoring, GHG emissions and associated information (fuel consumption, emission factors, heat and electricity production, and grid losses) are measured and analysed. During verification, the authenticity of the calculated emission reductions is assessed. This is done by a different DOE to that which performed the validation. The DOE ensures that the CERs are in accordance with the guidelines and conditions agreed upon in the initial validation. Certification is a written assurance by the DOE that, during the nominated time period, the project has achieved the emission reductions calculated. Project participants are notified of the certification outcome, and this includes a request to the EB to issue CERs equal to the verified amount. The EB issues CERs within 15 days of this request.

The ongoing costs apply over a crediting period. The crediting period is the time period chosen, prior to registration and stated in the PDD, by the project proponents over which CERs are claimed. A project can have either: a single 10-year crediting period; or a 7-year crediting period that can be renewed twice (subject to approval by the EB), giving a maximum of 21 years [1]. The process for renewal approval includes: an updated PDD using the latest approved baseline methodology, and validated by the DOE; a new letter of approval; and a request for renewal.

The crediting period begins once the project enters its operation phase, however its start can be delayed or brought forward by up to a couple of years (UNFCCC [11]—EB 52 Report, Annex 59). This altering of the start date can only be made once for each project.

UNEP [8–10] and others give typical transaction costs for small-scale (<15 MW) and large-scale (>15 MW) projects. Transaction costs vary according to the scale of the project. UNEP [8], p. 65 comments that 'project developers generally expect upfront transaction costs within the range of 5 to 7% of the net present value of the revenue or total transaction costs around 10–12% of the net present value of revenue'. Many of the transaction costs are fixed and therefore smaller-scale projects may have higher transaction costs per tonne of CO_2-e, making them less attractive to investors.

The registration fee paid is dependent on the amount of CERs generated, and is calculated according to a scale; the more CERs generated, the higher the rate per

CER. The registration fee is capped, and least-developed countries are exempt from paying this fee [1].

8.2.2.4 Revenue

Revenue in the following refers only to carbon-based revenue, and not to total project revenue (which would include the sale of the project end-product). Only carbon-based revenue is relevant in the following option calculations.

CERs are generally claimed on an annual basis within the crediting period, once verification and certification have been achieved. The CERs generated by the project may be sold. Their financial value equals the project CER output multiplied by the CER market price.

The market value of CERs during the (future) crediting period is crucial to establishing whether a project is worthwhile registering as a CDM project. Future carbon prices might be estimated based on historical performance as well as consideration of events that might impact the future price. UNEP ([8], p. 61) comments that 'the pricing of CER is highly speculative'. Consideration needs to be given to both the compulsory and voluntary carbon credit supply and demand, the various country emissions trading schemes, carbon legislation, economies and government policies around the world. At times, there might be an oversupply of carbon credits and the price of carbon might be low due to a disparity between demand and supply—the CER market conditions can shift.

Viability as a CDM project, in contrast to other project types, requires that the CDM-specific costs be less than the CER revenue generated.

8.2.3 The Relevant Options Analysis

8.2.3.1 Outline

The options analysis is based on the cash flows resulting from and including exercising the option (that is, claiming the carbon credits). At time i, $= T_S, T_S+1, \ldots, T_F$, let the revenue generated by the CERs claimed be Y_{i1}, and let the cost of claiming the carbon credits be Y_{i2}, where T_S and T_F are the start and finish times (years) of the crediting period. Y_{i1} and Y_{i2} are random variables, though in many cases only Y_{i1} will be a random variable. Then, with reference to Appendices 2.11.1 and 2.11.2, $X_i = Y_{i1} - Y_{i2}$. This can be enlarged if the costs and revenues occur over more than one time period; these other costs and revenues can be discounted to i. The rest of the equations in Appendix 2.11.2 follow. And the option value is calculated from Eq. (1.1) [4].

8.2.3.2 Case Examples

The calculations are demonstrated on two projects—one on hydropower, the other on wind power. All relevant information regarding costs and crediting periods are from the respective published PDDs. Interest rates used are those applicable to the country during the relevant crediting periods. To this, optimistic, most likely and pessimistic CER price estimates are made using the best available information. The project PDDs use crediting periods of 10 years.

8.3 Results

8.3.1 Case Example—Hydropower

8.3.1.1 Description

The case example covers a hydropower project, supplemented with additional material necessary for option calculations. The project is canal-based, utilising the flows and head available from a diversion channel. It is classified as a small-scale project due to its total installed capacity of 2 MW. The project generates electricity through sustainable means and is not anticipated to cause any harm to the environment, while contributing to emissions abatement.

The project has a fixed crediting period of 10 years. The project is anticipated to have zero project and leakage emissions, thus emission reductions are equal to the baseline emissions. PDD baseline calculations show that the hydro plant generates 10356 MWh per annum with an emissions factor of 0.9032 t CO_2-e/MWh, giving baseline emissions as 9357 t CO_2-e per annum.

8.3.1.2 Upfront/Pre-operational Costs

The upfront costs are estimated as \$80k (based on the project's PDD, or estimated—UNEP [8–10]), being comprised of:

- Search costs and Project Design Document (PDD)—\$40k;
- Baseline methodology and estimate of emissions in absence of the project—\$15k;
- Approval (by designated national authority) costs—\$12.5k;
- Validation costs (project eligibility regarding CDM requirements)—\$12.5k; and
- Registration—\$0.

8.3.1.3 Annual Ongoing/Operational Costs

The ongoing annual costs are estimated as $15k–$20k (UNEP [8–10]), being comprised of:

- Monitoring costs, including reporting—$5k;
- Verification costs—$5k (with an extra $5k, once only, at the start of the crediting period);
- Certification (assurance by DOE of the emission reductions)—$5k; and
- Adaptation fee (used to fund climate adaptation assistance)—2% of CERs generated.

This gives for annual total cost, Y_{i2}, $E[Y_{i2}]$ =$15k (with $20k in year 1 of the crediting period)

These costs are assumed here to be relatively certain over time, that is $Var[Y_{i2}] = 0$.

The last bullet point is taken care of in the following calculations by only using 98% of the CERS generated as income, rather than 100%.

8.3.1.4 Annual Revenue

In establishing CER price estimates, historical data and carbon market reports are available commercially. For example purposes, pessimistic, most likely and optimistic estimates are taken as $1.00/t, $2.50/t and $8.00/t respectively, and apply throughout the whole crediting period. This gives an expected value and variance of the price of $3.17, and ($1.17)^2 respectively.

The PDD gives estimated annual project CER output (emission reductions) as 9357 t. Based on Balatbat et al. [2], and for example purposes, pessimistic, most likely and optimistic estimates are assumed to be 5000 t, 7500 t and 10,000 t respectively. This gives an expected value and variance of the CER yearly output of 7500 t, and (833 t)^2 respectively.

Other values for CER prices and output, including different values for each year, might be assumed by different investors. For example, the variance in the CER price might be assumed to increase with time in order to reflect future uncertainty in a better way. However the form of the calculations, exampled here, remains the same.

For independence of CER price and output, this gives for annual revenue, Y_{i1},

$$E[Y_{i1}] = 3.17 \times 7500 = \$23.8\text{k}$$

$$Var[Y_{i1}] = 3.17^2 \times 833^2 + 7500^2 \times 1.17^2 + 1.17^2 \times 833^2 = (\$9.2\text{k})^2$$

8.3.1.5 Present Worth and Option Value

With expected values and variances in hand for the ongoing annual costs and annual revenue, these may be discounted to the start of the project using Appendix 2.11.2. The discount rate given in the PDD is 9.41% per annum. Assuming that this incorporates some allowance for uncertainty, a 7% per annum interest rate is used here for example purposes. Other rates might be assumed by different investors; however the form of the calculations, exampled here, remains the same. Equation (1.1) is used to calculate the option value. Table 8.1 gives the calculations for the end of each year of the crediting period, relative to the start of the crediting period.

The additional value (here 11.06% over all years), obtained by allowing options in each year, is demonstrated in Fig. 8.1. The added value comes from capping any downside resulting from low carbon prices and/or low project CER output, but taking advantage of any upside resulting from high carbon prices and/or high project CER

Table 8.1 Hydropower case example; yearly option values compared with deterministic present worth (E[PW]); StDev is standard deviation

Crediting period (year)	E[PW] ($k)	StDev[PW] ($\sqrt{\text{Var[PW]}}$) ($k)	Option value ($k)
1	3.55	8.60	5.45
2	7.69	8.04	8.31
3	7.18	7.51	7.76
4	6.71	7.02	7.25
5	6.27	6.56	6.82
6	5.86	6.13	6.35
7	5.48	5.73	5.92
8	5.12	5.35	5.57
9	4.79	5.00	5.19
10	4.47	4.68	4.82
Sum	57.13		63.45
% increase			11.06

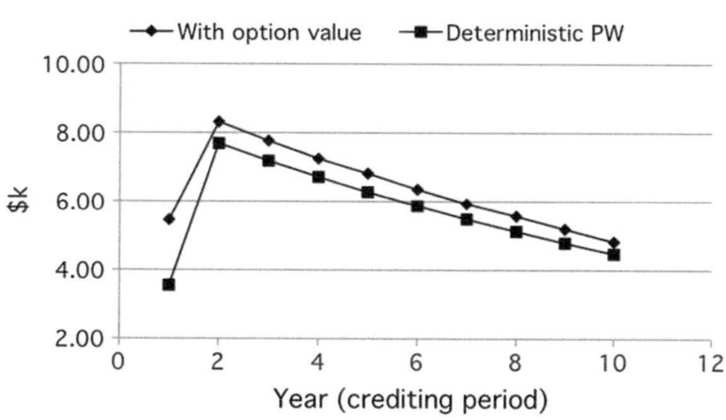

Fig. 8.1 Hydropower case example; additional value, obtained by allowing yearly options, is the difference between the two plots

output. Comparing \$57.13k or \$63.45k with the upfront cost of \$80k, the project is not viable from a CDM viewpoint; however allowing options decreases the loss to the investor.

8.3.2 Case Example—Wind Power

8.3.2.1 Description

The second case example used to demonstrate the value of allowing options is based on a bundled wind power project. It involves the installation of 5 wind turbine generators, of total capacity 7.5 MW and anticipated to generate 17838 MWh per annum. As a bundled project, the project only needs a single validation report and a single certification report.

Generated power is sold to a local electricity authority. The project contributes to emission reductions through replacing the burning of fossil fuels. The generators are assumed to not use any fuel and therefore the project emissions are taken as zero.

The PDD uses a 10-year crediting period.

8.3.2.2 Upfront/Pre-operational Costs

The upfront costs are estimated to total \$73.94k (based on the project's PDD, or estimated using UNEP [8–10]), being made up as follows:

- Search costs and Project Design Document (PDD)—\$13.5k;
- Baseline methodology—\$10k;
- Approval—\$5k;
- Validation costs and services—\$42k;
- Registration—\$1.69k; and
- Monitoring plan—\$1.75k

8.3.2.3 Annual Ongoing/Operational Costs

The ongoing annual costs are estimated as \$20k [8–10], being comprised as follows:

- Monitoring—\$10k;
- Verification—\$5k; and
- Certification—\$5k.

As well, there is a UN adaptation fee calculated at 2% of CER revenue.

This gives for annual total cost, Y_{i2},

$$E[Y_{i2}] = \$20k$$

These costs are assumed to be relatively certain over time, that is $Var[Y_{i2}] = 0$.

8.3.2.4 Annual Revenue

The PDD gives annual CER output (emission reductions) of 16001 t CO_2-e for years 1 to 8, dropping to 15841 and 15683 t CO_2-e for years 9 and 10 respectively. Using the reasoning of the previous hydropower case, and for example purposes, yearly pessimistic, most likely and optimistic estimates are assumed to be 8000 t, 12,000 t and 16,000 t respectively. This gives an expected value and variance of yearly CER output of 12,000 t, and $(1333\ t)^2$ respectively.

CER price estimates, for example purposes, are taken as the same as the hydropower case study.

Other values for CER prices and CER output, including different values for each year, might be assumed by different investors. However the form of the calculations, exampled here, remains the same.

For independence of CER price and output, this gives for annual revenue, Y_{i1},

$$E[Y_{i1}] = 3.17 \times 12000 = \$38.0k$$

$$Var[Y_{i1}] = 3.17^2 \times 1333^2 + 12000^2 \times 1.17^2 + 1.17^2 \times 1333^2 = (\$14.7k)^2$$

8.3.2.5 Present Worth and Option Value

The PDD uses a discount rate of 8.1706% per annum. Assuming that this includes some allowance for uncertainty, the following calculations use a lower value of 7% per annum, for example purposes. Other rates might be assumed by different investors; however the form of the calculations, exampled here, remains the same.

Appendix 2.11.2 is used to calculate present worths and Eq. (1.1) to calculate option values. Table 8.2 gives the calculations for the end of each year of the crediting period, relative to the start of the crediting period.

The additional value (here 2.58% over all years), obtained by allowing options in each year, is demonstrated in Fig. 8.2. The added value comes from capping any downside resulting from low carbon prices and/or low project CER output, but taking advantage of any upside resulting from high carbon prices and/or high output. Comparing \$126.42k or \$129.68k with the upfront cost of \$73.94k, the project is viable from a CDM viewpoint; however allowing options increases the viability, albeit here only slightly, to the investor.

Table 8.2 Wind power case example; yearly option values compared with deterministic present worth (E[PW])

Crediting period (year)	E[PW] ($k)	StDev[PW] ($\sqrt{\text{Var[PW]}}$) ($k)	Option value ($k)
1	16.82	13.74	17.20
2	15.72	12.84	16.13
3	14.69	12.00	15.02
4	13.73	11.21	14.08
5	12.83	10.48	13.19
6	11.99	9.80	12.34
7	11.21	9.15	11.46
8	10.48	8.56	10.72
9	9.79	8.00	10.10
10	9.15	7.47	9.44
Sum	126.42		129.68
% increase			2.58

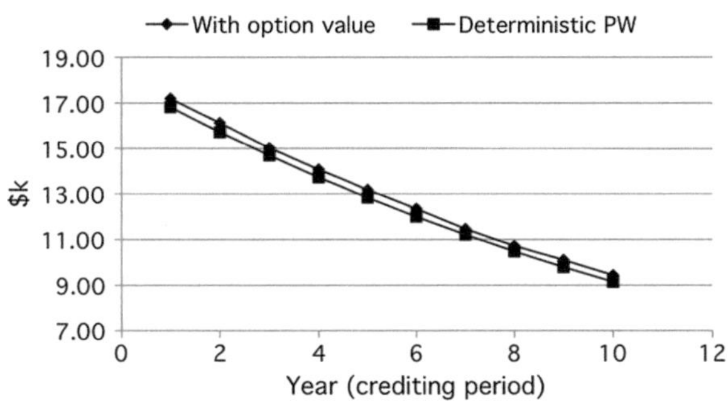

Fig. 8.2 Wind power case example; additional value, obtained by allowing yearly options, is the difference between the two plots

8.4 Discussion

8.4.1 General Comments

In amplification of the above analysis and case example calculations, the following comments are given:

i. Whether it is worthwhile to seek CDM registration for a project requires a comparison of the upfront/pre-operational costs and on-going/operational costs with the revenues gained. Additional to this, allowing options improves the viability. Options may convert an unviable project into one that is viable, while they increase the viability of an already viable project. It is seen in the case examples presented, that even when the overall project deterministic present worth is negative, options still have a value, albeit small in some cases. Incorporating options always leads to a value greater than the equivalent deterministic present worth of any investment. This added value comes from having the right but not the obligation to claim or not claim carbon credits at any time. Numerically, this was demonstrated in the case examples. Only when the feasibility, Φ, has a value of 1 does the option value not add anything to the project viability. However, in this case the investment is already sufficiently attractive without any consideration of options.

ii. The calculations rely on forecasting future CER prices, project CER output and CDM-based costs, and their variability. Any means may be used to do this, for example, for CER prices, using a combination of time series models, experience, and knowledge of factors (supply/demand, …) affecting prices (Chap. 2). Forecast estimates inherently contain uncertainty. CDM projects have uncertainty in future CER price, project CER output and CDM-based costs. It is this uncertainty which gives an option its value. The greater the uncertainty (as measured by variances), the greater the option value. In the case examples, the costs were assumed fixed; however, the option value will increase if costs also contain uncertainty (based on costs and revenue being largely independent).

iii. The general conclusions of the chapter are not sensitive to differing assumptions in crediting period, interest rate, and uncertainty and magnitude of CER price, project CER output, and compliance cost. The magnitudes of the calculated option values differ, but not the overall conclusion that allowing options increases the attractiveness of CDM projects.

iv. The transaction costs per CER for large-scale projects are smaller than for small-scale projects where the transaction costs can be quite significant. Hence it is anticipated that having recourse to options will benefit small-scale projects more than large-scale projects. Any simplification of procedures and standardization of reporting would reduce transaction costs and increase the option values.

v. The case example analyses were based on the possibility of exercising options on a yearly basis. The analysis is essentially the same should options be allowed to apply to several consecutive years treated as one; revenues for each year are summed and discounted, while on-going costs may increase but be fewer in number, and would also be discounted. It is seen that allowing options in any form increases the attractiveness of a CDM project.

vi. The inclusion of options does not affect CDM financial additionality, which essentially only requires that the project not be viable without CER revenue. The inclusion of options only affects viability from the project investor's viewpoint.

vii. It is seen that allowing options comes at no detriment to any stakeholder. It is a win-win situation—global emissions are reduced further by making more CDM projects viable, while project proponents potentially could gain financially, or lose less. Opportunity costs are no different to, or are less than the current CDM situation, where options are not permitted.

8.4.2 Policy

The chapter suggests that CDM rules should be modified in order to improve the attractiveness of CDM projects. The increased attractiveness, in turn, should lead to more CDM projects, contributing to greater global emission reductions. The modification suggested is that options be allowed.

Currently CDM rules allow for minor accelerating or delaying the claim for CERs at the start of the crediting period.

This is a form of option; at the start of the crediting period, if the carbon price is low (high) and/or the project output is low (high), and the cost of CER verification and issuance is greater (less) than the CER value, the project proponent can defer (bring forward) making a claim for CERs up to 2 (1) years. However, this only applies at the start of the crediting period, and is once off. This chapter is proposing that the option to claim or not claim CERs should be available throughout the entire crediting period, at the discretion of the project proponents, and not just at the start of the crediting period.

8.4.3 Related Issue

An issue, related to the one discussed in this chapter, is the choice of crediting period. A project can either have a single 10-year crediting period, or a 7-year crediting period that can be renewed twice, giving a total of 21 years [1]. If a 7-year crediting period is renewed, it must first be approved by the EB, and this requires, among other things, an updated PDD with the latest version approved baseline methodology and validated by the DOE.

The choice of an initial 7-year crediting period, and the subsequent renewal or non-renewal for a following 7 or 14 years, might be seen as a sequential option, that is an option to discontinue or continue as a CDM project at the end of year 7, and later possibly at the end of year 14. The analysis of the option occurring at the end of year 7 (assuming renewal only for only one more 7-year period) is calculated in terms of the anticipated carbon-based costs (CDM operational costs) and anticipated carbon-based revenues (based on forecast CER prices and forecast project CER output) for years 8–14. All these costs and revenue contain uncertainty. This option value then needs to be compared with its 'premium', namely the (additional) cost of preparing an updated PDD at the end of year 7. The option value would need to exceed the additional PDD cost in order to proceed as a CDM project. That is, there could be value in having flexibility to either discontinue or continue a project at the end of year 7, dependent on anticipated costs and anticipated revenue at year 8 and over the following years up to year 14.

Similarly, if the choice at the end of year 7 is that of continuing CDM registration, there exists an option at the end of year 14. The value of this option is calculated in the same way as for the option at the end of year 7, where now the cash flows relate to anticipated carbon-based costs and anticipated carbon-based revenues for years 15–21. This option value then needs to be compared with its 'premium', namely the (additional) cost of preparing an updated PDD at year 15. Again, there could be value in having flexibility to either discontinue or continue a project at the end of year 14, dependent on anticipated costs and anticipated revenue at year 15 and over the following years up to year 21.

The total option value of having options at the end of both years 14 and 7 is evaluated by working backwards in time. The option value at the end of year 14 is calculated based on the cash flows in years 15–21. This option value, together the additional PDD cost at year 15 and the cash flows in years 8–14, give the option value at the end of year 7. This option value, together the additional PDD cost at year 8 and the cash flows in years 1–7, give the option value at year 0. This is the total option value and is compared with the up-front CDM cost in order to establish CDM viability.

This interpretation of the 3 by 7-year sequential option adding value to CDM projects is original.

8.5 Closure

The chapter proposed that options be allowed within CDM projects. Introducing options will require a CDM rule change. It is suggested that rules similar to CDM rules already existing could be introduced. Such rules would not impact any other existing CDM rules. This idea of allowing options within CDM projects is original.

An option gives the project proponent the right but not the obligation to gain should the going price of CERs and/or the project CER output generated be favourable. The uncertainty associated with future CER prices, project CER output and costs are turned to the investor's favour. Two CDM projects, used as case examples, demonstrated the benefits of allowing options.

Allowing options within CDM rules improves the attractiveness of CDM projects. In turn, this should lead to more CDM projects contributing to sustainability via more emission reductions. It is seen that allowing options comes at no detriment to any stakeholder. It is a win-win situation—global emissions are reduced further by making more CDM projects viable, while project proponents potentially could gain financially, or lose less.

8.6 Extensions

Related projects. The chapter's conclusions carry over to related initiatives, such as the Carbon Farming Initiative (CFI) [6]. CFI projects can also be demonstrated to be more attractive if options were allowed.

Future research. It is conceivable that, if available, options might only be exercised on many projects when the carbon price is high, and not when the carbon price is low, even though exercising is also dependent on individual project CER output. This could alter the supply-demand balance of carbon credits, and in turn affect the carbon price. This supply-demand-price relationship could be explored.

References

1. Baker and McKenzie (2015) CDM Rulebook. http://www.cdmrulebook.org. Accessed 31 Jan 2015
2. Balatbat MCA, Findlay E, Carmichael DG (2012) Performance risk associated with renewable energy CDM projects. J Manag Eng 28(1):51–58
3. Carmichael DG (2014) Infrastructure investment: an engineering perspective. CRC Press, Taylor and Francis, London
4. Carmichael DG, Ballouz JJ, Balatbat MCA (2015) Improving the attractiveness of CDM Projects through allowing and incorporating options. Energy Policy 86:784–791
5. Carmichael DG, Lea KA, Balatbat MCA (2016) The financial additionality and viability of CDM projects allowing for uncertainty. Environ Dev Sustain 18(1):129–141
6. Department of the Environment (2015) About the carbon farming initiative, Department of the Environment. http://www.environment.gov.au/climate-change/emissions-reduction-fund/cfi/about. Accessed 31 Jan 2015
7. Schneider L (2007) Is the CDM fulfilling its environmental and sustainable development objectives? An evaluation of the CDM and options for improvement. Öko-Institut, Berlin, Germany
8. UNEP (United Nations Environment Program) (2003) CDM information and guidebook, ed M-K Lee, UNEP Riso Centre, Roskilde, Denmark
9. UNEP (United Nations Environment Program) (2007) Guide to financing CDM projects, CD4CDM project. UNEP Riso Centre, Roskilde, Denmark
10. UNEP (United Nations Environment Program) (2011) CDM information and guidebook, 3rd edn. UNEP Riso Centre, Roskilde, Denmark
11. UNFCCC (United Nations Framework Convention on Climate Change) (2015a) What is the CDM. http://cdm.unfccc.int/about/index.html. Accessed 31 Jan 2015
12. UNFCCC (United Nations Framework Convention on Climate Change) (2015b) CDM insights. http://cdm.unfccc.int/Statistics/Public/CDMinsights/index.html. Accessed 31 Jan 2015

Chapter 9
Convertible Contracts

9.1 Introduction

On projects, as work is completed and new information comes to hand, the scope becomes better defined. Equivalently, uncertainty in the scope decreases. It would seem reasonable then that the contractual arrangements on a project should also progress in line with this decreased uncertainty, rather than staying constant from project start to project end. Information asymmetry, whereby the project contractor is better informed of the work than the project owner, and contractor self-interest leading to possible opportunistic behaviour, also suggest the need to tailor contractual terms to a project's situation in order that the owner's interests and the contractor's interests better align. It could also be anticipated that tailoring a contract to a project would be more cost-effective than having a rigid structure applicable throughout the project duration. Such tailoring also enables an adjustment of project risk, and may possibly assist in the reduction of disputes. In line with these potential advantages, the chapter explores the role and viability of changing contract terms, as a project progresses, and offers an original analysis in this respect.

For definiteness, the chapter concentrates on construction projects and contracts with conversion between payment types, but the chapter's approach generally applies to all contracts and all terms within contracts. General discussion that is broader than construction and payment types, that is, covering contracts and terms generally, is given in a supplemental form. The broader usage of the word 'terms' incorporates payment types as a particular example.

Contract payment types may be classified as either fixed price contracts (including lump sum, schedule of rates or unit price, and guaranteed maximum price—GMP) or as prime cost contracts (including all the cost reimbursable varieties). Each payment type gives a different risk for the owner and the contractor, and each has different applicability dependent on the degree of information known about the project work.

D. G. Carmichael, *Future-proofing—Valuing Adaptability, Flexibility, Convertibility and Options*, Management in the Built Environment,
https://doi.org/10.1007/978-981-15-0723-6_9

Having the ability (flexibility) to discretionarily change (convert, switch) contract payment types can be interpreted as an option, and its value can be established using the book's approach. The holder of the option to convert is given the right, but not the obligation, to change payment types. It could be anticipated that such a right would have a cost or a premium (possibly, but not necessarily up front), and so it is necessary to value the option in order to establish what is a reasonable cost to pay for the right. The definite conversion, that is, one that is contractually devised to happen with certainty as opposed to a discretionary conversion, can be shown to have a lower value, and in some cases a negative value. Conversion might be anticipated to occur at the transition point between defined project stages [3], because of the different knowledge, skills and resources that are required for these stages. But, conversion need not be restricted to this timing.

The option value is established here by looking at the project net cost resulting from the change, and in particular the expected value and variance of the net cost, and then discounting these to present worth values. The net cost is the difference between the cost of continuing under the existing payment type and the cost that will occur under the new payment type. The option value follows from knowing the present worth distribution fitted to these discounted values. This, in conjunction with any cost or premium (possibly, but not necessarily up front), establishes whether flexibility embedded within a contract is worthwhile or not from a commercial viewpoint. Legal issues could involve the termination of the first contract, formation of the second contract, novation, and a variation; these are discussed below. The chapter does not discuss potential agency issues [1, 5–7].

The structure of the chapter is as follows. Both commercial and legal implications are covered. Firstly, the background to convertible contracts is outlined. The characteristics of convertible contracts are then examined, along with how the flexibility associated with convertibility is valued. Finally, associated legal matters are discussed, followed by a case example on building refurbishment, a summary of the approach from the point of view of the contracting parties, and implementation issues. Reference terminology is to owner and contractor in the following as the contracting parties, though the material is applicable for contractual dealings between any two parties.

The chapter provides an original approach to valuing convertible contracts. Current literature does not deal directly with the matter addressed. The chapter, for definiteness, concentrates on construction contracts with conversion between payment types, but the chapter's approach applies to all contracts and all terms within contracts. The chapter will be of interest to researchers and practitioners concerned with optimizing project outcomes through contract design. The theory and results are independent of any particular conditions of contract. Rather, mention is only made to contract terms which generally apply across all conditions of contract; any conditions of contract can be modified to include a conversion capability. The chapter does not discuss the strengths and weaknesses of different payment types or different conditions of contract [2].

9.2 Background

The chapter draws its background from contract payment types, options and convertible contracts. Contract payment types, their place of usage and associated risk issues are established knowledge and are not therefore repeated here (see, for example, [2]).

The concept of having a 'convertible' entity is not new. For example, it has existed for some time in the finance industry in bonds, securities and venture capital transactions, and in insurance policies and adjustable rate mortgages, allowing for changes in contractual terms such as the length of insurance or the type of interest rate applicable.

However, the introduction of the concept to contracts on projects is still in the development stage, with little work published to date. Despite this, there is a belief that convertible contracts offer advantages when compared to more traditional ways of contracting. Contract conversion has been used in practice, but without any formal analysis or an attempt to value it or frame it within an options analysis.

Incorporating options into contracts gives an owner an opportunity to mitigate its costs. The owner is able to hedge against cost escalation, obtain potential gains, and prevent post-contract-formation opportunism. A convertible contract prevents the contractor from behaving inefficiently and from taking unfair advantage of any such opportunism.

In a construction context, the convertible contract idea has a cousin in Early Contractor Involvement (ECI) project delivery. In ECI, the contractor for the first stage (involving full or part design, and possibly other activities such as project risk mitigation) may be selected through a non-price competitive process, making the relationship between the contractor and the owner important, as it is in all prime cost contracts. Whilst there is a form of ECI that uses one contractor for the design stage and then seeks tenders from others to undertake the construction, in convertible contracts there is only one contractor involved.

9.3 Convertible Construction Contracts

The term 'convertible contract' is used as an umbrella term to describe a flexible form of contract. It allows an owner and a contractor to agree on one set of terms initially and later agree different terms depending on the circumstances at the time, or upon reaching defined events. Once the terms have changed, they are regarded as having been 'converted'. In the context of construction contracts, these terms may relate to a range of things, including payment type, conditions of work, the programming/scheduling of work, or the project scope. The contract is tailored to the specific conditions applicable to the project at the time of the conversion. Accordingly, the use of a convertible contract should be advantageous.

It is unlikely that all the terms of a contract are subject to change. Those terms that are flexible would be agreed upon by the contracting parties and clearly stated in the contract as terms which may be later converted. The extent to which these terms may be converted would also be explicitly stated, and are essentially contingent upon the agreement between the owner and the contractor. Alternatively, contracting parties may choose to agree on the extent of conversion upon initial engagement, without the need for later amendments or the introduction of new terms. At the time of conversion, price negotiation between the parties may be necessary. This chapter centres on converting payment types.

Incorporating the flexibility to change contract payment types can be seen as an option, and its valuation follows the options analysis of this book. To be considered worthwhile for the owner to incorporate an option into a contract, the value of the option (to the owner) needs to exceed any cost for incorporating the option. To be acceptable to the contractor, the contractor also needs to gain from converting or switching contract payment types.

Generally, it would be envisaged that payment switching would take place from a prime cost payment type to a fixed price payment type; that is, the project commences under a prime cost payment type and later converts to a fixed price payment type. This is so, because at the start of a project the scope may be ill-defined and hence suitable for a prime cost type payment, but as the project progresses and its scope becomes better defined, a fixed price type payment would be more appropriate. Within the two payment types, any of the varieties could be considered. The switch occurs at the point where the remaining scope can be reasonably accurately defined and priced, and only if the switch is worthwhile (to the option holder) at the time. The decrease in scope uncertainty, together with the switch of payment types, lead to different risks to each party.

The following discussion refers to giving the option right to the owner. Since the benefit lies with the owner, inclusion of a conversion clause in the contract is contingent on agreement with the contractor, and providing the contractor with appropriate compensation, or the contractor gaining in some way. However, if the contractor perceives working under a fixed price arrangement better than under a prime cost arrangement (in which case, both the owner and contractor stand to gain), this payment might be small. The payment might be viewed as an incentive to agree to a convertible contract. In addition, the contractor is afforded the opportunity to demonstrate goodwill to the owner and to build on its reputation and future relationship. It would also be possible to give the option exercising right to the contractor, or to be by joint agreement of the parties at the time of exercising. However, giving the owner the right would appear to provide the owner with the best project outcome, including the elimination of agency issues with the contractor; hence it is this situation which this chapter considers.

9.4 Method of Analysis

Discussion in the following refers to one switch or conversion in the lifetime of the project; for multiple switches, the situation repeats.

Common to all convertible contracts are:

- An exercise or switching time, T, chosen on the basis of having enough project information to estimate a fixed price finishing cost with reasonable accuracy. The time T would be contractually defined, either approximately or exactly, based on prior knowledge of the project and its timeframe, and known to both parties. The owner can exercise the right to switch at time T.
- There are both costs and benefits flowing to both the owner and contractor at and following time T. These costs and benefits will be different to those had no switch taken place.
- There is a cost to the owner for having the option right. This might be manifested in a payment or benefit to the contractor at any time in the project.

The only suitable way of analysing an option of the form considered here is by using the book's approach; traditional financial option analogies are regarded as being inapplicable. Only costs relating to exercising the option are considered, that is, costs at or after the time of exercising (T). The costs for the second payment type are subtracted from the costs for the first payment type (assuming that it continues in the absence of switching). All costs are probabilistic, and are discounted to give the probabilistic present worth. The option value then follows on knowing the present worth distribution.

Cost estimates, whether approximate, preliminary or detailed, are assembled in the usual way. Cost estimates and their variability could be obtained from historical data, or built up from components, including direct costs (directly related to work items), indirect costs (time-varying and one-off—not identified directly with work items), business overheads and profit. Contingencies are incorporated into the estimate variability in the component costs, and would not be a separate component as in a deterministic estimate. In terms of the component approach, the uncertainty in each component can be established in a number of ways, for example as presented in Chap. 2. Correlations between the cost components are also needed, and these might best be estimated based on knowledge of the particular components and their paired relationships, rather than through any involved mathematical analysis. For example, the greater the volume of work (direct cost), the greater the time-varying cost (indirect cost), while one-off costs such as site disengagement may not be affected; profit and business overheads might be added as a percentage, the quantum of which might vary with market conditions and competition. If the total estimate is being built up, say, through monthly costs, then some thought will also need to be given to the correlation between monthly cost components—generally high correlation could be anticipated between similar cost components close in time, but reducing with time apart and as the nature of the work alters. For the intent and accuracy of the analysis, it would not be necessary to look at sub-component cost estimate correlation. It would also only be necessary to broadly categorize the degree of correlation, for example as high, medium, low or no correlation.

Cash flows are looked at from the viewpoint of the option holder, such that something favourable to the holder is taken as positive and something unfavourable to the holder is taken as negative. This leads to looking at the difference between the future

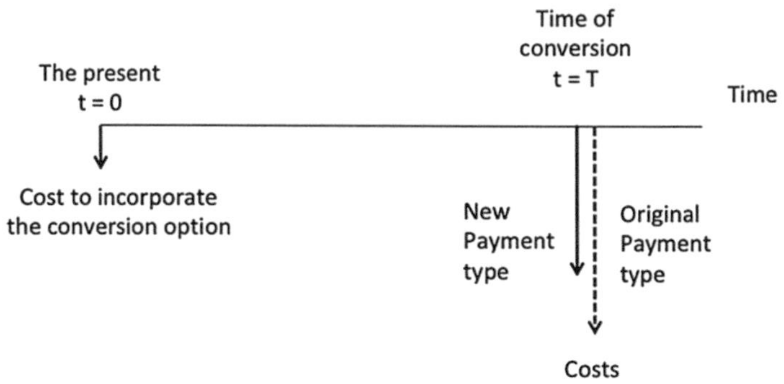

Fig. 9.1 Cash flows for comparative analysis

cost if the original contract type is continued at and beyond T and the future cost if the new contract type is adopted at and beyond T.

For costs at and beyond T, let Y_{T1} be the (total) original payment type cost estimate assuming that the original payment type continues (that is, in the absence of switching), and Y_{T2} the (total) new payment type (replacing the original payment type, with switching occurring) cost estimate (Fig. 9.1, where costs at and beyond T have been discounted to T). Close to perfect correlation may apply between these two cost estimates. At T,

$$X_T = Y_{T1} - Y_{T2}$$

The expressions in Appendices 2.11.1, 2.11.2 and 2.11.3 follow.

The value of having the flexibility to convert can then be established using Eq. (1.1), namely, convertibility value = ΦM. The payment type is only switched at time T if it is worthwhile to do so.

This convertibility value might be shared between owner and contractor by negotiation. Based on the origin of Eq. (1.1), a definite conversion (leading to E[PW]) has a lower value than the convertibility value, because of the negative part of the PW distribution, and can even take negative values.

It follows from Appendices 2.11.2 and 2.11.3 that the greater the uncertainty in the cost of the project, the greater the value of this flexibility. Hence, the value of this flexibility (the value of convertibility) decreases as the project scope becomes better defined, and as estimates are being made for activities less distant into the future, that is, as the project progresses. The cost uncertainty derives from a shortfall in scope definition, and trying to foresee the future and predict future events. There are multiple sources of uncertainty in construction, for example, latent conditions, adverse weather and potential inaccuracies in any forecasted costs. However, it is because of this uncertainty that the flexibility has a value; with determinism (that is, the remaining project cost is known with certainty), the flexibility has no value.

The time at which the option is to be exercised might be pre-determined in the contract or may be variable. The analysis is similar for both cases. If a variable time, then the best time to exercise the option could be established by enumeration—doing the same option calculation for different T values and selecting the best outcome. The option is only exercised if, at the time, it is worthwhile to do so. That is, it is only exercised if there is a return to the owner by changing contract payment types. For example, if guaranteed maximum cost (GMC) is being used to start the project off, and the total project cost, if switching occurs, would exceed the maximum cost guaranteed by the contractor, the option may not be exercised, because staying with GMC and not switching may yield the lowest project cost.

9.5 Case Example

The following case example relates to the upgrading of a residential building through part demolition of the building (primarily the roof structure) and the addition, among other things, of an additional storey [4].

The building demolition stage introduces many uncertainties as to the existing building condition, while the subsequent construction stage (new building work) is relatively routine building construction. Accordingly, a cost reimbursable contract is appropriate for the demolition phase, and a lump sum contract is appropriate for the new building work.

The switch between payment types (namely, cost reimbursable to lump sum) occurs following the end of the demolition stage, equivalently the start of the construction stage. The time at which the switch occurs, that is, the time of option exercising T, is the time planned for the end of the demolition phase. Here, the planned switch time T = 1 year.

Cost estimates are assembled from all component costs. This is a well-established practice and so is not given here. Rather, the calculations are worked here in terms of total estimated costs. Only those costs at or beyond the time of exercising, T, are relevant in the calculations, not the total project cost.

For the new building work, the (total) estimates are:

- Under a cost reimbursable contract—\$680k most likely estimate, with optimistic and pessimistic estimates of ±25%. Using the PERT expressions relating optimistic, most likely and pessimistic estimates to expected value and variance (Chap. 2), this gives $E[Y_{T1}] = \$680.0k$ and $Var[Y_{T1}] = (\$56.7k)^2$, where Y_{T1} refers to cost reimbursable values.
- Under a lump sum contract—\$650k most likely estimate, with optimistic and pessimistic estimates of ± 15%. Using the PERT expressions relating optimistic, most likely and pessimistic estimates to expected value and variance (Chap. 2), this gives $E[Y_{T2}] = \$650.0k$ and $Var[Y_{T2}] = (\$32.5k)^2$, where Y_{T2} refers to lump sum values.

Since values for Y_{T1} and Y_{T2} are based on similar estimating principles, it is reasonable to assume that Y_{T1} and Y_{T2} are close to being perfectly correlated. That

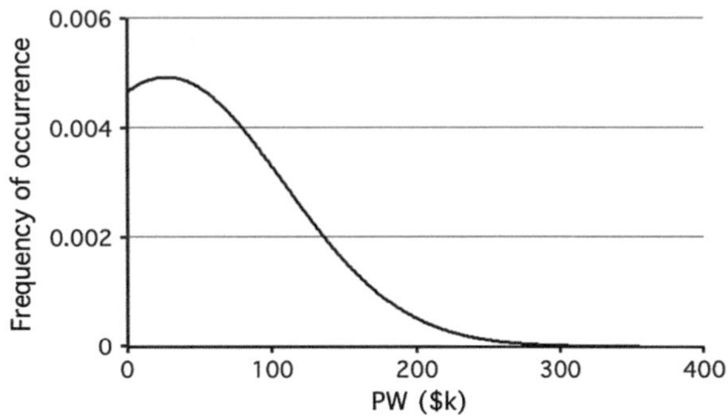

Fig. 9.2 Example—part probability distribution for PW (PW upside)

is, as estimates for Y_{T1} increase (decrease) so too should estimates of Y_{T2} increase (decrease).

For an assumed 10% per annum interest rate, the expected value and variance of the present worth can now be calculated by discounting the difference between Y_{T1} and Y_{T2} to the present (Appendix 2.11.2), E[PW] = \$27.3k and Var[PW] = (\$81.1k)^2.

To demonstrate the next calculation, though this is not necessary in practice, the probability distribution for PW can be drawn. Assuming a normal distribution fitted to the values of E[PW] and Var[PW], gives Fig. 9.2. Figure 9.2 shows the 'PW upside'.

Equation (1.1) requires two values, namely Φ and M. Here, $\Phi = 0.618$ and $M = 73.0$. Then, from Eq. (1.1) the convertibility value = 0.618×73.0 = \$45.1k. This is approximately 7% of the contract sum (for the new building work).

Comparison. Incorporating a definite switch in payment types within the contract (compared with the option to switch as just calculated), leads to a lower value (namely E[PW] = \$27.3k or 4% of the contract sum).

The switch may or may not occur depending on the circumstances at time T. The switch only occurs if it is in the owner's interests to do so at T.

The financial analysis has a number of input variables, the most central of which are the uncertainty in the cost estimates and the interest rate. Increasing the uncertainty in the cost estimates increases the convertibility value. However, the conclusions remain essentially the same with different interest rates, primarily because any cost estimates for each payment type are being discounted similarly.

The value associated with an option to switch will decrease with time as the uncertainty in the cost estimates reduces and the scope becomes better defined. The option to switch provides a hedge against risk associated with cost uncertainty, and its value decreases as the risk decreases.

9.6 Summary Approach

The approach to be adopted in establishing the financial viability of incorporating convertibility in a contract is as follows:

1. The time, T, at which conversion may take place is estimated.
2. The costs for both (original and new) payment types for the project at and beyond T are estimated. These may be in the form of optimistic, most likely and pessimistic values, from which expected values and variances of costs can be calculated. Cost correlations, if developing the estimates from components, are also estimated.
3. These costs are discounted to the present day and the differences between the two sets of costs lead to an expected value and variance of PW (Appendix 2.11.2). Based on this, a distribution can be fitted to PW.
4. Equation (1.1) is used to calculate the convertibility value.
5. Negotiation may then proceed between the owner and the contractor as to how this value (less any cost or premium necessary in order to establish the option) might be distributed between the parties.

Optimal (in the sense of greatest reward) timing of the conversion would clearly be of interest to the parties in order that they gain as much as possible. This can be established though enumeration. That is, different exercising times are considered in Step 1 above, and the option value for each calculated, with the best of all calculated values noted. However, project considerations, for example related to the type of work or resourcing and not monetary value alone, may dictate the 'best' time to switch.

Where a project is composed of different work items, each covered by a different payment type, it is possible that any of these work item payments could be separately subject to being changed during the project.

9.7 Legal Matters

A contract is a legally enforceable agreement, and hence some thought needs to be given to legal matters.

The contractual power to change would be conferred by a 'conversion clause'. Such a clause would allow the owner the right, but not the obligation, to change the type of contractor remuneration as detailed within the clause, after work had begun. The exercising of the option would not be brought about as a surprise or with any punitive or threatening purpose. The ability to switch along with mention of the timing would be contractually agreed.

The following discussion is within a common law framework, and focuses on particular points which are applicable to convertible contracts.

(i) An options clause. An appropriate clause would allow the owner an option to switch from a defined prime cost remuneration type to a defined fixed price remuneration type. The clause should also stipulate something about how the fixed price is to be obtained, and the time(s) at which the option may be exercised.

(ii) Novation or variation. There are two possible ways or scenarios in which the switch between contractor remuneration types might be looked at legally. Firstly, where exercising the option amounts to a fundamental alteration to the terms of the first contract such that a new contract is required to be formed; and secondly, where exercising the option amounts to a mere variation to the terms of the first contract such that no new contract is required, but rather a variation of the first contract is required to be enforced.

In the first scenario, a novation results, because the first contract is required to be formally terminated and discharged and the second contract is required to be properly formed to replace it. The parties expressly agree to a specific way in which the first contract will be terminated, namely by exercising the option in a defined way. The result of this is, in essence, two separate contracts each with their respective terms, obligations and rights. Since only changes to the contractor's remuneration type are involved, it is anticipated that all other terms from the first contract will be mirrored in the second contract. Being commercial dealings, the elements of a contract—agreement, consideration, legal intention, capacity, …, would be apparent in the second contract as in the first contract. The option cost to the owner might also play a role in being regarded as consideration in the discharge of the first contract and/or the formation of the second contract. Another possibility is that two contracts exist from the start of the project, with only one being applicable at any time during the project.

In the second scenario, the first contract would require drafting in such a way as to convey the change in contractor remuneration as a minor adjustment to the terms of the first contract and as a subsequent agreement. Here, reference is to a variation of a contract (as distinct from variation of the work), which generally involves adding, omitting or altering of contractual terms. This requires that the parties agree that exercising the option only changes the terms of the first contract. No new contract is required to be formed, but rather a subsequent agreement is created. The consideration for this subsequent agreement could be the option cost, or the change in or forgoing obligations (for example, payments and work) for both parties. Since no new contract is being formed, it is anticipated that all terms, other than the change to the contractor's remuneration, would continue to apply once exercising has taken place.

Both scenarios—novation and variation—are equally valid at law. It is thus possible to structure a contractual arrangement to suit how the parties choose to be engaged. However, the variation approach may offer a more seamless transition between remuneration types.

9.8 Implementation Issues

With convertible contracts, the main implementation issues, which need addressing, are the following:

- There is a lack of industry familiarity with the concept of convertibility. Generally, human nature is such that something new may initially be distrusted. However, looking beyond any initial reaction, the parties should be able to assess any associated risks and see the benefits.
- The time of conversion would need to be reasonably well bounded and the form of conversion would need to be reasonably clear in the contract, in order not to spring any surprises on the contractor.
- The owner and contractor would need to negotiate the sharing of any gain from having a conversion capability in the contract.
- Negotiated bounds would need to be placed on the lump sum in order to discourage contractors from de-railing the conversion, after having agreed to its inclusion. This might be in the form of an agreed a priori estimate qualified in terms of the associated uncertainty (plus/minus a percentage). Only the form of the fixed price arrangement can be agreed upon; the specific monetary value remains to be negotiated at the time exercising is contemplated. The final agreed lump sum price would be based on a full knowledge of the risks explored in the prime cost stage, but would be by negotiation between the owner and contractor.
- There would need to be mechanisms established for: paying for work started but not yet complete and paid for under the cost reimbursable payment type; recognizing work paid for but not yet complete under the cost reimbursable payment type; and for clearly delineating between the scope covered under the cost reimbursable payment and the scope covered under the lump sum payment. Work not covered by the latter scope would sensibly be paid for on a cost reimbursable basis.

The calculation steps given above lead to a monetary value for having the option to switch within a contract. However, the deeper messages, independent of any numerical calculations, are:

- The value associated with having an option to switch will always be greater than the value associated with a definite switch (one which is contractually stated will occur). This is so, because some switches can lead to a loss; in such cases the option is not exercised, capping any loss at zero. A contractually prescribed definite switch can, in some cases, have a negative value.
- Having the option to switch will always be more attractive than having a contract where switching is not foreseen. Having an option to switch always has a value, even if small in some cases.
- The option value derives from the presence of uncertainty in the cost estimates. The greater the uncertainty in the cost estimates, the greater will this option value be. Conversely, if the there is no uncertainty in the cost estimates, having an option gives no added value. As a project progresses and any uncertainty decreases, the option value decreases.

9.9 Closure

The idea of having flexible contracts is not new, but heretofore a rational analysis has been missing. This chapter gives the analysis as well as an examination of contracts with conversion capabilities. The chapter, for definiteness, concentrated on construction contracts with conversion between payment types, but the chapter's approach applies to all contracts and all terms within contracts. Current literature does not deal directly with the matter addressed in this chapter. Related published work examines conversion in the context of post contract formation opportunism exhibited by contractors, and market underlyings that exhibit price volatility in contradistinction with projects.

The chapter examined the commercial and legal issues surrounding convertible contracts. It is found that both owner and contractor could benefit from such contracts. From the owner's perspective, benefits include hedging against potential financial losses, and increased contractor efficiency. For the contractor, advantages arise from the opportunity to gain from the conversion and to show good faith, thereby increasing its potential for future relationship contracting. The more efficiently the contractor works under a lump sum arrangement compared to a cost reimbursable arrangement, the higher the value of convertibility becomes to the contractor.

Having the right, but not the obligation, to convert the payment type within a contract part way through a project has a value. This flexibility value can be readily calculated. How and when this is shared between the contracting parties would be negotiable between the parties. The direct share to the contractor might be viewed as part of a premium paid by the owner for having flexibility in the contract.

Being a right but not an obligation, the switch may or not be done depending on the circumstances at time T. Clearly, the owner would only switch if it is worthwhile to do so at T.

The summary findings on alternative contractual approaches are:

- The value associated with having an option to switch will always be greater than the value associated with a definite switch (one which is contractually stated will occur).
- Having the option to switch will always be more attractive than having a contract where switching is not foreseen.
- The greater the uncertainty in the cost estimates, the greater will this option value be. As a project progresses and any uncertainty decreases, the option value decreases.

9.10 Extensions

The chapter examined conversion between payment types, but equally the approach is applicable to conversions of other sorts including delivery methods, contract clauses, subcontracts, project expansion and project abandonment.

The chapter examined total project conversion. However, it would also be possible to do part conversions, that is, some subprojects may be converted while other subprojects may not. While the analysis would remain the same as that given in this chapter, research could examine any nuanced implementation issues.

References

1. Boukendour S (2007) Preventing post-contractual opportunism by an option to switch from one contract to another. Construction Management and Economics 25:723–727
2. Carmichael DG (2000) Contracts and international project management. A. A. Balkema, Rotterdam
3. Carmichael DG (2004) Project management framework. A. A. Balkema, Rotterdam
4. Carmichael DG, Karantonis JP (2015) Construction contracts with conversion capability: a way forward. Journal of Financial Management of Property and Construction 20(2):132–146
5. Hosseinian SM, Carmichael DG (2014a) Optimization in the development of target contracts, chapter 15. In: Xu H, Wang X (ed) Optimization and control methods in industrial engineering and construction. Springer Engineering, Berlin, pp 259–296
6. Hosseinian SM, Carmichael DG (2014) Optimal sharing arrangement for multiple project outcomes. Journal of Financial Management of Property and Construction 19(3):264–280
7. Hosseinian SM, Carmichael DG (2014) Optimum sharing in project delivery methods. International Journal of Engineering Management and Economics 4(2):151–175

Chapter 10
Financial Options

10.1 Introduction

Appendix 1.5 of Chap. 1 shows the relationship between a Black-Scholes European call option and the book's approach leading to the Carmichael equation, Eq. (1.1) [2].

Using the notation of Chap. 2 leading to the appendices in Chap. 2, then the conversion of traditional financial options terminology can be readily stated. A risk free value for interest rate r is used.

A European option is exercised on a specified date; an American option can be exercised at any time before or on the specified date. The exercise price of an option is stated in the options contract, and might be negotiated between the buyer and the seller. These are in the form of contracts between two parties.

Options have maximum losses capped at their premiums.

European call option

Let S_T be the stock price at T. The exercise price, K, at T is deterministic. Then,

$$Y_{T1} = S_T$$

$$Y_{T2} = -K$$

$$Y_{Tk} = 0 \qquad k = 3, 4, \ldots, m$$

$$X_i = 0 \qquad i \neq T$$

A put option reverses the signs of Y_{T1} and Y_{T2}.

D. G. Carmichael, *Future-proofing—Valuing Adaptability, Flexibility, Convertibility and Options*, Management in the Built Environment,
https://doi.org/10.1007/978-981-15-0723-6_10

Put-call parity

The put option is obtained by looking at the other tail of the present worth distribution to that of a call option, and is evaluated in the same way as a call option, in both cases looking at the cash flows from the option holder's viewpoint, namely a cash inflow being positively viewed and a cash outflow being negatively viewed, irrespective of any accounting conventions.

Put-call parity can be shown to hold [1].

American option (call and put)

Let the time to exercising the option follow a probability distribution. The choice of this distribution is at the discretion of the person doing the calculations, and would reflect the investor's belief in the likelihood of exercising within the time T; uniform or triangular distributions might be suitable. Let p_t be the probability that the exercise takes place in period t, t = 0, 1, 2, …, T, and let $PW_{(t)}$ be the associated present worth. Then,

$$E[PW] = \sum_{t=0}^{T} p_t E[PW_{(t)}]$$

$$Var[PW] = E[PW^2] - \{E[PW]\}^2$$

Comment

Dividends, transaction costs and taxes can be incorporated, in both the European and American option cases, by regarding these as additional cash flows, and they may be probabilistic.

An implied variance may be calculated similarly to implied volatility. The Greeks can be obtained through conventional sensitivity analysis.

An assumption is required on the interest/discount rate. It would appear reasonable to use the cost of capital for the investor or the opportunity cost of capital unadjusted for most risk, because the uncertainty in the cash flows is accounted for in the variance terms. As well, any uncertainty caused by distant time is included in the estimation of the optimistic, most likely and pessimistic values for the cash flows.

The exercise cost is considered to be just another cash flow (positive or negative as the case may be depending on the type of option) in the year(s) that exercising takes place. For example, buying implies a negative cash flow; selling implies a positive cash flow.

10.2 Example Calculations

A demonstration of how the proposed method may be implemented on both real and financial options follows.

Real option example

Let the exercise cost be \$125,000 at year 5, and assume that it can generate expected positive cash flows of \$25,000 for each of years 5 to 10. Let the interest rate be 10% per annum. The deterministic present worth of such an investment is −\$3,250, implying that the investment is not worthwhile using such a criterion. However variability produces an upside.

1. Estimates are required of the expected values and variances of each of the positive cash flows. Assume that the investor estimates that optimistic and pessimistic values for these cash flows are ±50% either side of the expected values, which are here assumed to be also most likely values. Using the earlier expressions given for calculating expected values and variances from optimistic, most likely and pessimistic estimates, the expected value and variance of each yearly positive cash flow are \$25,000 and 17,360 ($\$^2$) respectively.
2. Discounting the exercise cost (negative) and positive cash flows to the present,

$$E[PW] = -\$3{,}250 \text{ (as in the deterministic case)}$$
$$Var[PW] = 26{,}280 \ (\$^2)$$

3. Calculate the feasibility Φ, the probability that the present worth is positive. This is the area under the positive part of the present worth distribution. Calculate the mean of the present worth upside. These calculations can be done using either the equations for a normal distribution to get accurate values, or for an approximate value, divide the upside part of the present worth distribution into vertical strips and calculate its area and centroid as a structural engineer would calculate for a member cross section (Chap. 2). Here $\Phi = 0.27$, and upside mean = \$3,110, giving,

$$OV = \$830$$

This is an estimate of the option value. (With a volatility of 3.1% derived using Appendix 2.11.4, Black-Scholes gives an option value of \$1420. As a proportion of the exercise cost of \$125,000, the difference is 0.47%.)

Financial option example

Consider an example. Let the exercise cost/price of the stock be \$10.20 at year 2. Let the interest rate be 10% per annum.

1. Estimates are required of the expected value and variance of the value of the stock at year 2. Based on past performance of the stock, the investor estimates/forecasts that the optimistic, most likely and pessimistic values of the stock will be \$13.86,

$9.90 and $5.94 respectively. Using the earlier expressions given for calculating expected values and variances from optimistic, most likely and pessimistic estimates, the expected value and variance of the value of the stock at year 2 are $9.90 and 1.74 ($\$^2$) respectively.

The deterministic present worth of the investment is −$0.25, implying that the investment is not worthwhile using such a criterion. However the variability produces an upside.

2. Discounting the exercise cost (negative) and stock value (positive) at year 2 to the present,

$$E[PW] = -\$0.25 \text{ (as in the deterministic case)}$$

$$Var[PW] = 1.19\ (\$^2)$$

3. Calculate the feasibility Φ, the probability that the present worth is positive. Calculate the mean of the present worth upside. (See point 3 above.) Here $\Phi = 0.41$, and upside mean = $0.78, giving,

$$OV = \$0.33$$

This is an estimate of the option value. (With a volatility of 9.4% derived using Appendix 2.11.4, and S_0 obtained by discounting $9.90 at 10% over 2 years, Black-Scholes gives an option value of $0.36. As a proportion of the exercise cost/price of $10.20, the difference is 0.29%.)

The above calculations are based on a normal distribution assumption for present worth. However any other distribution considered a suitable model for present worth can be used if preferred by the investor.

10.3 Appendix: Carbon Options

Carbon options are not covered in this book. However, using the book's approach, they are treated no differently to energy options in Chap. 11, or property options in Chap. 12.

References

1. Carmichael DG (2014) Infrastructure investment: an engineering perspective. CRC Press, Taylor and Francis, London
2. Carmichael DG, Hersh AM, Parasu P (2011) Real options estimate using probabilistic present worth analysis. Eng Econ 56(4):295–320

Chapter 11
Energy Options

11.1 Introduction

Economic, supply and demand, and external factors affect the price of energy. The price exhibits uncertainty, high volatility, seasonality, and spikes. One way of dealing with risk [1] associated with energy prices, is through hedging with energy options. The value of energy options depends on the underlying price of energy. In line with options generally, an energy option grants the option holder the right to buy and/or sell energy at a pre-agreed price. Modified Black-Scholes and binomial and trinomial lattices and related methods can be used to value the options in usual markets.

However, in non-usual or atypical markets, including those where something unforeseen happens in energy production or supply, or there are shifts in economic conditions or external factors, such pricing tools may not apply. By comparison, the book's approach to valuing options is applicable to atypical markets, as well as usual markets. Energy prices can be estimated as outlined in Chap. 2, and can incorporate experience, historical performance, knowledge of the market, and knowledge of anticipated shifts in economies or external factors. Predetermined time series describing the evolution of energy prices over time can also be used to inform, but the book's method does not rely on such information alone.

This chapter explores different types of energy options—plain call and put options, spark spread options, swing options, and options associated with callable and putable forwards. The chapter will be of interest to anyone involved with derivatives in energy markets.

11.2 Option Uses

Energy options may be used for hedging or limiting risk associated with fluctuating prices. Stakeholders can lock-in a specific future energy price in order to protect themselves from high or low energy prices caused by unfavourable

D. G. Carmichael, *Future-proofing—Valuing Adaptability, Flexibility, Convertibility and Options*, Management in the Built Environment,
https://doi.org/10.1007/978-981-15-0723-6_11

market fluctuations. Options enable energy retailers to limit their exposure; if energy prices alter greatly, retailers can exercise their options and transact at a contracted exercise price, either buying from or selling back to the producer.

The energy industry is said to have utilised options for many decades through terms and conditions of trade, however the US and UK restructuring and deregulation in the 1990s appears to have promoted the more recent usage of options [2].

In line with options generally, options can also be used as a speculative tool, whereby losses are capped because of the non-obligatory nature of their exercising. In the energy market, options usually have short to medium maturity times ranging from months to a few years [2].

11.3 Call and Put Options

Plain call and put options are traded based on an agreement between the retailer and producer. Call options may be used for example by energy retailers to hedge against a rise in energy prices. Energy retailers may also be able to sell back energy to the producers, creating a put option that enables the retailers to hedge against price drops when energy is to be returned. The option is only exercised when the energy price drops below the exercise price. Call and put options are also used for speculative purposes and are traded on a stock exchange.

Plain options enable retailers to meet supply obligations without serious harm to profit margins during periods of price increases, and when unexpected events such as price spikes and generator outages occur. Electricity producers can also hedge against events such as a generator outage through holding options.

The book's approach relies on making an informed estimate of the energy price (at the time of exercising T), S_T, and its uncertainty. The informed estimate can incorporate knowledge of the market, national and international economies, seasonality, experience, education, historical performance, and 'gut feel'; it does not have to accord with any pre-conceived time series modelling of energy prices, but would obviously be informed by such modelling. The discussion in Chap. 2 on estimating is relevant. With a second order moment analysis in mind, estimates are made of the moments, $E[S_T]$ and $Var[S_T]$. Both S_T and the exercise price, K, are discounted to time 0 to give the present worth, PW. Then, for interest rate r,

$$E[PW] = \frac{E[S_T] - K}{(1+r)^T} \qquad Var[PW] = \frac{Var[S_T]}{(1+r)^{2T}}$$

This is a particular case of that outlined in Appendices 2.11.1, 2.11.2 and 2.11.3. With E[PW] and Var[PW] in hand, any two-parameter probability distribution of choice can be fitted to PW. The option value, OV, follows from Eq. (1.1). The procedure outlined in Sect. 2.3 is followed, but now specifically using estimates of $E[S_T]$ and $Var[S_T]$.

The put option is handled similarly (Chap. 10).

Example and comparison with Black-Scholes. A numerical example is used to demonstrate the calculations involved as well as the closeness of the results of the book's approach and Black-Scholes.

Assume estimates of the energy price at T = 1 year, obtained via estimating optimistic, most likely and pessimistic values, are $E[S_T] = \$11.05$ and $Var[S_T] = (\$2.81)^2$. Prices are per unit of energy. These are discounted along with the exercise price of \$13.00 to give $E[PW] = -\$1.77$, and $Var[PW] = (\$2.55)^2$, for an interest rate of 0.10 per annum. Using a normal distribution for convenience (but the person doing the calculations is free to use any appropriate distribution), OV = \$0.37.

To compare with Black-Scholes, the equivalent volatility, σ, using Appendix 2.11.4, is 0.25. Discounting S_T, gives $S_0 = \$10.05$, leading to a Black-Scholes option value of \$0.42, a difference of \$0.05 or 0.4% as a proportion of the exercise price. Assumed PW distributions, other than a normal distribution, could make this difference smaller.

11.4 Spark Spread Options

Spark-spread call options grant the option holder the right, but not an obligation, to pay $H \times S_{GT}$ on exercising at T, and in return receive one unit of electricity; H is the fuel heat rate and S_{GT} is the unit price of fuel at T, the time of exercising.

Spark spread calls allow energy producers to hedge against unfavourable price movements and uncertainty in energy production inputs. The payoff relates to the difference between the energy price, S_{ET}, and $H \times S_{GT}$. Spark spread put options, which grant the option holders the right but not an obligation to pay for one unit of electricity and receive $H \times S_{GT}$ at T, appear less common and more used for speculative purposes.

Where the market is atypical, Black-Scholes and similar approaches would generally not work. This book's approach, as with plain call and put options, can deal with any market irregularities or regularities, and also incorporate uncertainty in the heat rate and exercise price.

Using notation similar to [3], a spark spread call option written on fuel G at a heat rate H, gives the option holder the right but not the obligation to pay H times the unit price of fuel G at T, and receive one unit of electricity.

Let S_{ET} and S_{GT} be the unit prices of electricity and fuel at time T, respectively. This chapter's approach relies on making informed estimates of S_{ET}, H and S_{GT}, and their uncertainty, as outlined earlier. In line with a second order moment analysis, estimates are made of the moments $E[S_{ET}]$, $Var[S_{ET}]$, E[H], Var[H], $E[S_{GT}]$ and $Var[S_{GT}]$. Let $K = HS_{GT}$ be the exercise price. Then for H and S_{GT} independent,

$$E[K] = E[H]E[S_{GT}]$$

$$Var[K] = E^2[H]Var[S_{GT}] + E^2[S_{GT}]Var[H] + Var[H]Var[S_{GT}]$$

and,

$$E[PW] = \frac{E[S_{ET}] - E[K]}{(1 + r)^T}$$

$$Var[PW] = \frac{1}{(1 + r)^{2T}}[Var[S_{ET}] + Var[K] - 2Cov[S_{ET}, K]]$$

For any of the above variables being deterministic, then its variance is zero. The option valuation then proceeds as for the plain call option. The put option also proceeds similarly.

11.5 Forwards

Forwards are contracts between a producer and a retailer, and contain agreed set terms and conditions for trading energy at some point in the future. Forwards are non-standardised derivatives and are negotiated between the producers and retailers. Forwards are valued based on past prices. A forward contract holds no value itself; its value is that of the underlying at the time of maturity. In energy derivatives, forwards may not be used alone, but rather include derivatives.

A common hedging technique, which combines the use of forwards and options, is callable and putable forwards. A callable forward involves one party to the contract to buy one forward and sell one call option with a predetermined exercise price. The other party holds the opposite position. In putable forwards, one party buys one forward contract and a sells a put option. The other party holds the opposite position. The effect of these trades is to hedge the risk associated with price shifts. Such options can be evaluated as for the plain call and put options developed above.

11.6 Swing Options

Swing options involve multiple exercise rights before the maturity date, but packaged as one. The volume of energy, which is traded, lies within a defined range. The exercise price of the contract is fixed throughout the life of the contract and is set at the beginning of each time period.

Swing options may be bundled with forward contracts. A commonly used method for pricing swing options is via dynamic programming and trinomial trees. The option valuation method outlined in this chapter, rather than say trinomial trees, could be used in conjunction with dynamic programming.

11.7 Closure

Energy prices are uncertain, and energy options provide one way of dealing with the associated risk. In atypical markets, including those where something unforeseen happens in energy production or supply, or there are shifts in economic conditions or external factors, traditional option pricing tools may not apply. This chapter showed how energy options can be valued in such atypical markets as well as regular markets. It offers a practical way of valuing such options. It relies on future probabilistic estimates of energy prices, which can be made by any means available to the person doing the calculations, and can incorporate experience, knowledge of the market, and knowledge of anticipated shifts in economies or external factors. The approach allows the person doing the calculations to intuitively adjust the value of the option by incorporating any information available. It does not rely on any predetermined time series describing the evolution of energy prices over time, though any existing knowledge on time series, trend lines and related forecast thinking can be used to inform price estimates.

References

1. Carmichael DG (2016) Risk—a commentary. Civ Eng Environ Syst 33(3):177–198
2. Deng SJ, Oren SS (2006) Electricity derivatives and risk management. Energy 31(6–7):940–953
3. Deng SJ, Johnson B, Sogomonian A (2001) Exotic electricity options and the valuation of electricity generation and transmission assets. Decis Support Syst 30(3):383–392

Chapter 12
Real Estate Options

12.1 Introduction

Economic, political and other external factors influence the value of real estate or property and the uncertainty in this value. One way of dealing with this uncertainty and associated risk is hedging through options. An option grants the option holder the right, but not the obligation, to typically buy or sell something in the future. Black-Scholes [3], binomial lattices and equivalent Monte Carlo simulation are commonly used pricing tools to value these options.

However, in atypical markets, where property price fluctuations and price uncertainty are anticipated to not follow any usual patterns, such pricing tools may not apply [6]. This chapter shows how options in such atypical markets or atypical situations can be valued according to the book's approach, which also applies in regular markets. The approach is straightforward to implement, intuitive to understand and requires low mathematical sophistication. Future cash flow estimates are informed by experience, knowledge and education related to the market, together with any available analysis (Chap. 2). Although no time series or trend lines describing the time evolution of property prices are used, if available they can act as information supplemental to the experience, knowledge and education of the person doing the calculations.

The approach is demonstrated on a range of different types of real estate options, and compared with Black-Scholes. The range includes property (purchases, sales, expansion, delay and abandonment), stock exchange indices, mortgages (prepayment, default/close, change), and leases.

The approach is one of interpreting an option as giving rise to equivalent future cash flows, discounting these to the present day using probabilistic present worth, and interpreting the resulting present worth distribution. Cash flows not associated with exercising are not considered, but are nevertheless important from any overall investment viability viewpoint. The cash flows are interpreted from the option

D. G. Carmichael, *Future-proofing—Valuing Adaptability, Flexibility, Convertibility and Options*, Management in the Built Environment,
https://doi.org/10.1007/978-981-15-0723-6_12

holder's viewpoint as in usual property investments. Thus there is no need to distinguish between expand or contract style options. For typical options, the results agree with Black-Scholes, but the approach is less restrictive on assumptions than Black-Scholes. The structure behind the approach is consistent with Black-Scholes. However, any information available to the person doing the calculations, qualitative or quantitative, can contribute to making informed estimates or forecasts of future cash flows. The approach is also generally applicable to all real estate options, irrespective of the underlying—real property, indices, rentals or other.

The outline of this chapter is as follows. Firstly, some background on real estate options and the book's approach is given. A range of real estate options is then shown how they can be interpreted in terms of the book's approach. A conclusion follows. All options presented in this chapter are valued from the viewpoint of the option holder. The chapter does not look at the effects on other parties, surrounding circumstances, people behaviour, or industry issues. Taxation and duties applicable on property transactions are country dependent, and mention of these is omitted from the following, but would be included in an overall viability calculation of any transaction. The chapter will be of interest to anyone involved with options in real estate.

12.2 Background

The body of knowledge on financial options is typically borrowed when dealing with real estate options. In real estate, the underlying may be property (including land) belonging to one of the parties to the option, may be some index, or may relate to mortgages and leases.

Options protect against losses beyond the premium paid, but are rewarded for gains. Options are used for hedging, speculation and conducting arbitrage transactions. Greater uncertainty in the underlying price gives greater value to the option. Options in real estate may be used, for example, to hedge against fluctuating prices, at property buying time to assist gaining finance, or in anticipation of land rezoning. For a property purchaser, an option can put a property on hold up until the expiry date of that option, thereby providing time to secure appropriate finance, insurance and permits before committing to the purchase.

With property as the underlying, the options tend to be traded directly between the parties, but perhaps overseen by agents. The option contracts are non-standard and are tailored specifically to the parties' needs, for example in terms of time to maturity; not being standardised leads to an absence of contract writing guidelines, regulations and details. American or European option styles can be used. Factors affecting the value of the underlying, and hence the value of the option premium include the property's size, location, age and condition, as well as property demand and price forecasts. A characteristic of property is that it is relatively illiquid when compared to stocks and bonds, and characteristics of property options are that they are private transactions, infrequently traded, with higher fees, information asymmetry and many local markets. This leads to decreased numbers of arbitrageurs in real estate option trading [6]. Options contracts can be sold on a secondary market to another interested buyer, if available, should the original option holder not wish to proceed with the purchase or not be able to get the necessary finance, insurance or permits. The sale terms on the secondary market are negotiable, dependent on the amount of time remaining until expiry of the option, demand for the property and how prices may have shifted since the original option contract was initiated.

Instead of potentially taking ownership of property, stock exchanges have real estate indices enabling investors to speculate using options. This facilitates an affordable way of investing in the property market; option contracts can be at relatively low cost with low trading and brokerage costs, and there are reduced borrowing requirements and expenses compared to taking ownership of the property with all its associated costs.

Within mortgages a number of options exist; these include options related to prepayment, default/close and change. Within leases, also a number of options exist; these include subleasing, lease cancellation, lease renewals, property purchase and revenue sharing.

12.3 Option Valuation

Black-Scholes, binomial lattices and Monte Carlo simulation equivalents, and their variations, are commonly accepted ways of valuing options. These require assumptions on the movement (such as Geometric Brownian motion) in the underlying price, assumptions on the distribution of prices (for example log normally distributed), and other assumptions such as a constant risk-free interest rate. Adjustments are made to suit particular uses. Such methods are not followed here in favour of the book's approach.

A typical call option involves paying a premium now, and in return having the option to purchase the property for an amount K (exercise price) at some time T in the future. Let the estimate of the property value (or underlying price) at time T be S_T. From the option holder's viewpoint, K can be regarded as a negative cash flow, and S_T a positive cash flow. The option is exercised if it is worthwhile (in the money), that is $S_T > K$. $S_T - K$ can be discounted to give a present worth, and the option value follows from Eq. (1.1).

Equation (1.1) provides a single way of valuing all options, whether call-style or put-style, whether related to property, indices, mortgages or leases. Later, the cash flows S_T and K are generalised to be cash flow equivalents, interpreted from the option holder's viewpoint. Favourable cash flows resulting from exercising the option are regarded as a cash inflow, while unfavourable cash flows are regarded as a cash outflow. Strict accounting conventions need not be used. Monte Carlo simulation or a second order moment analysis can be used to discount the cash flows (Chap. 2). This book uses the latter for convenience. It requires no assumptions to be made on the probability distributions of the cash flows, but rather works in terms of the moments, E[] and Var[], of the cash flows. Possible ways of estimating the expected values and variances of underlying prices or values are discussed in Chap. 2 and the next section.

Agreement between the book's approach and Black-Scholes is good. Section 12.5 gives a comparison example. Higher uncertainty in an underlying price or value leads to greater spread in the present worth distribution, and hence higher option values. Put-call parity can be shown to hold [4]. Exact agreement with Black-Scholes could not be anticipated because Black-Scholes and the book's approach are based on different assumptions, most notably (with the book's approach): volatility is replaced with variance and variance need not be constant over time; there are no required time series or probability distribution assumptions; the exercise price need not be deterministic or at one point in time; interest rates can vary over time and be different for the exercise price and the underlying price or value; and discounting can be continuous time or discrete time.

12.4 Estimates of Future Underlying Prices or Values

Estimating practices vary between people, and invariably involve combining knowledge of: the market; historical trends; industry data; and national and international economies and events. This is coupled with experience, education and 'gut feel' (a combination of logic and emotion). Any fitted time series models or trend functions available can be incorporated and to whatever extent desired, but will need modification in atypical markets, perhaps through qualitative manipulation, where the past is not repeating. There is no one single agreed method for estimating. What is required is a best estimate based on available information. Estimating is not dealt with in more detail in Chap. 2 All underlying prices and values—property prices, indices, rental costs etc.—contain uncertainty.

In applying Eq. (1.1), persons doing the calculations are free to adopt whatever method they like in order to estimate future underlying prices or values. In the following examples, estimates for the underlying prices or values at time T, S_T, are made by first estimating optimistic (a) most likely (b) and pessimistic (c) values, as in Chap. 2. But the book's approach can use estimates obtained by any method, including where uncertainty increases with time. The exercise price K is deterministic and is established from the options contract. Correlations between variables similarly can be estimated in any way considered suitable, but the results will be bound by assumptions of complete correlation and independence.

12.5 Applications

12.5.1 Outline

A range of example option valuations is given using the book's approach. The approach applies to both real property, indices, mortgages and leases without modification.

Black-Scholes, with alteration, and similar approaches, rely on future fluctuations in the underlying prices to be the same as the past (history repeating). Such approaches are well established and are not repeated here. Where the market is atypical, Black-Scholes and similar approaches would generally not work [6]. This book's approach however can deal with any market irregularities or regularities, and also incorporate uncertainty in all the variables.

Present worth, PW, when calculated through a second order moment analysis, requires estimates of the moments $E[S_T]$ and $Var[S_T]$. S_T and K are discounted to time 0, to give PW (Chap. 10 and Appendix 2.11.2),

$$E[PW] = \frac{E[S_T] - K}{(1+r)^T}$$

$$Var[PW] = \frac{Var[S_T]}{(1+r)^{2T}}$$

Knowing E[PW] and Var[PW] allows any two-parameter probability distribution for PW to be fitted. The distribution choice is discretionary. The option value, OV, then follows from Eq. (1.1). The process steps become (Chap. 2):

- Estimate $E[S_T]$ and $Var[S_T]$;
- Discount these together with K to give E[PW] and Var[PW];
- Fit any reasonable probability distribution to PW; and
- Calculate Φ and M, and hence OV.

The call option looks at one tail of the PW distribution. The put option looks at the other tail of the PW distribution. Whichever is being looked at, cash flows are interpreted from the option holder's viewpoint, namely a cash inflow being positively viewed and a cash outflow being negatively viewed, irrespective of any accounting conventions.

The analysis is done for any time of exercising, T. To establish the optimum time to exercise, repeated similar analyses are done for different times, the resulting option values compared, and the best selected by enumeration.

Normal distributions for present worth are used in the following examples, but any distribution thought appropriate can be used. An interest rate of 8% per annum (p.a.) is adopted.

The choice of interest rate and probability distribution for present worth is up to the person doing the calculations. The choice is based on whatever is thought to be the most suitable. With interest rate choice, it is noted that uncertainty in the property/index price is already incorporated, and so a choice approaching a 'risk free' rate might be preferred.

Some examples follow.

12.5.2 Property Purchase

Consider purchasing a property for redevelopment. The property is currently valued around \$1M but the buyer needs time to raise finance and get necessary development permits. The property value is anticipated to increase. To hedge against a potential price rise, the investor purchases a call option on the property at an exercise price of \$1M and an expiry date of one year. Optimistic, most likely and pessimistic estimates for the property in one year are \$1.05M, \$0.99M and \$0.95M respectively. That is, $E[S_T] = \$0.993333M$ and $Var[S_T] = (\$0.016667M)^2$. This gives $\Phi = 0.34$, $M = \$10{,}310$ and $OV = \$3{,}550$. The premium that the purchaser should be looking to pay will be something close to this OV.

In some option contracts, the final purchase price of the property might be reduced by the premium already paid by the option buyer, if the option is exercised. This does not affect the option value calculation, but does affect overall purchasing viability.

The above example involves usual market forces on property prices. A second example that uses the same calculations involves rezoning. Perhaps, rezoning of rural land might be imminent, and maybe dependent on a different political party in power. A developer may pay the land owner an option premium (call option) such that the developer has the right to purchase the land at some defined future time for a defined exercise price. Optimistic estimates (a) for S_T are now not based on historical trends, but rather require experience in similar previous situations, along with a sense of what the demand might be for a product that currently does not exist. Pessimistic estimates (c) will be based on the existing political party staying in power. Most likely estimates (b) will be based on the likelihood of a different political party. For example, with a probability p of the political party staying the same, $b = pa + (1 - p)c$.

The land owner receives the premium now. The developer's gain with a new political party is not limited. The developer's loss with the same political party is capped at the premium.

12.5.3 Property Sale

Options can also be used to hedge in the sale of property in the future. A property owner wishes to sell a property in one year. The property value may decrease. The owner purchases a put option on the property for an exercise price of \$1M and an expiry date of one year. Optimistic, most likely and pessimistic estimates for the property in one year are \$0.97M, \$1.01M and \$1.03M respectively. That is, $E[S_T] =$ \$0.993333M and $Var[S_T] = (\$0.01M)^2$. This gives $\Phi = 0.25$, M = \$5,540 and OV = \$1,400. The premium that the seller should be looking to pay, in order to lock in the \$1M sale price, will be something close to this OV. For example, for a premium of \$2000, a property value in one year greater than \$1.002M (option not exercised; property sold on the market) or less than \$0.998M (option exercised; property sold for \$1M) represents a gain to the seller. Between \$0.998M and \$1.002M sale price, the option holder suffers a loss, but the loss is capped at the breakeven values of \$1.002M and \$0.998M.

12.5.4 Property Expansion, Delay, Abandonment

Oppenheimer [6], among others, notes a number of real options existing with buildings and land—expansion, delayed development (wait), and abandonment. In a given situation, more than one of these options may exist, but on exercising one option, the other options are no longer present. The analysis of each of these options involves, as before, identifying the cash inflows and cash outflows resulting from exercising the option.

Expansion might be contemplated in order to exploit a new or growing market. Examples include upgrading buildings or follow-on housing project development. The cash outflows relate to the cost of expansion (including any possible disruption costs), and the cash inflows relate to the ensuing income generated by the expansion. All cash flows can contain uncertainty, and hence the assumption of a deterministic exercise price in Black-Scholes does not hold.

Abandonment and contraction are treated similarly. Contraction may involve selling real estate in anticipation of a downturn in demand. The relevant cash flows are the difference between what would have existed (assuming no contraction) and what new cash flows result from the contraction (usually both positive and negative), together with any sale income or disposal costs. If abandonment, there is a need to also include a one off demolition/disposal cost or residual value. Refer Chap. 3.

Delay or deferment of an option might occur in anticipation of some future event occurring or future market shifting, or future approval. The analysis of delays is no different to any of the previous options except that a delay may affect (increase or decrease) the magnitudes of the relevant cash flows. The presence of a delay may have its own cost. The option value may improve or decrease with a delay depending on the particular situation. The optimum delay time can be established by doing analyses for different exercising times and selecting the best result from these. Refer Chap. 3.

12.5.5 Stock Exchange Indices

Real estate performance is tracked with indices on stock exchanges. Options with these indices as underlyings are an alternative way to invest in real estate, and require relatively low capital compared with the outright purchase of property or stock. The exercise style is European, that is exercising upon maturity. Maturity may be quarterly and up to four quarters or a year total [1, 2]. In a rising market, an index call option may be bought, while in a falling market an index put option might be bought.

An example of the use of a real estate index option for hedging is that of an individual who wants to buy a house in one year's time but is anticipating that house prices will rise at a faster rate than the individual's income and savings. To counter this, the individual buys a call option on a real estate index, which has a high correlation with house prices. Another example is that of a homeowner who wants to sell in one year and would like to hedge against a decline in house prices. To counter this, the homeowner buys a put option on a real estate index. The hedging reduces exposure brought about by adverse price movements in associated assets.

Consider an example of an investor who buys a call option on a REIT (real estate investment trust) index. The exercise price is 1025 with a one-year maturity. Optimistic, most likely and pessimistic estimates are 1050, 1020 and 1000 respectively (at one year). That is, $E[S_T] = 1021.7$ and $Var[S_T] = (8.3)^2$. This gives $\Phi = 0.34$, $M = 5.13$ and $OV = 1.8$. The premium that the investor should be looking to pay would be considered relative to this OV—a premium lower than this OV would be desired by the investor. The discussion on gains and losses follows the earlier call option treatment; the loss is limited to the premium paid. Using the Appendix 2.11.4 expression for a volatility–variance relationship, Black-Scholes gives an option value of 3.0, which is about 0.1% different as a proportion of the exercise price. Assuming a present worth distribution other than a normal distribution, or assuming another volatility–variance relationship, could make this comparison closer.

12.5.6 Mortgages—Prepayment, Default/Close, Change

A number of authors have written on options that exist within mortgages. Some example options that exist within mortgages and mentioned by these authors relate to prepayment, default/close and change. These authors also discuss the circumstances when such options might be exercised and associated ramifications. The value of these options does not appear to be something that is explicitly calculated by the borrower, however the method given here can be used. At any particular time, T, the borrower looks at its effective cash flow position. The effective cash flow position at T is the present worth of all equivalent cash flows in and cash flows out, at and beyond T, discounted to time T. The net equivalent cash flow at T is then discounted to 0 to give PW, and the option value is calculated as above.

The options to the borrower for prepayment, default/close and change cannot exist at the one time. Exercising one option prevents the others from being exercised at the same time. Each option will have a different worth to the borrower, and it seems reasonable that the borrower would select the option which is best for it. Sensibly, an option is only exercised when it is 'in the money', but borrowers may have non-mortgage reasons for exercising.

All analysis is from the borrower's perspective—the borrower is the holder of the option. The lender's perspective can be informed by looking at this. Most publications tend to be from the lender's perspective in trying to establish the value of a mortgage in the presence of these options.

For dealing with options within mortgages over N years, the equations in Appendices 2.11.1, 2.11.2 and 2.11.3 can be used, and are a generalisation of the earlier Sect. 12.5.1 equations. For the option, equivalent cash inflows and cash outflows, denoted Y_{1i} and Y_{2i} respectively, will exist at and beyond T, namely for years $i = T$, $T + 1, \ldots, n$. Y_{1i} (in place of S_T) and Y_{2i} (in place of K) can be random variables. In any year i, the net cash flow is $X_i = Y_{1i} - Y_{2i}$. That is, $E[X_i] = E[Y_{1i}] - [Y_{2i}]$ and $Var[X_i] = Var[Y_{1i}] + Var[Y_{2i}] - 2\rho_{12}\sqrt{Var[Y_{1i}]}\sqrt{Var[Y_{2i}]}$, where ρ_{12} is the correlation between Y_{1i} and Y_{2i}. The present worth of X_i, $PW_i = \frac{X_i}{(1+r_i)^i}$. Allowing for the interest rate r_i to also be a random variable and to vary with year i, this leads to the results in Appendix 2.11.3.

Appendix 2.11.3 is a generalisation of Appendix 2.11.2. The use of Eq. (1.1) follows. This is the option value when viewed at time $i = 0$. As time progresses, updated option values can be calculated by discounting the cash flows to the current time i. It is noted that this book's method allows for non-deterministic exercise price and non-deterministic interest rate, unlike traditional financial options methods.

Prepayment

Prepayment refers to the ability of a borrower to prepay any remaining loan balance during the loan period, and can be interpreted as a call option [6]. The borrower in return takes ownership of the property. Prepayment may be done in conjunction with external or internal refinancing. In a market where the differential—property value minus mortgage value—increases with time, the likelihood of exercising the option would increase. Holding the option may be of no consequence to the borrower however, because the borrower may not have access to the prepayment amount, and this may be the case for many borrowers. A lender would not be aware of this, and hence would not know at any time whether the borrower was likely to exercise the option.

Cash inflow—property value, Y_{1T}; $Y_{1i} = 0$, $i = T + 1, T + 2, \ldots, n$

Cash outflow—mortgage value and transaction cost, Y_{2T}; $Y_{2i} = 0$, $i = T + 1, T + 2, \ldots, n$

Default/close

Borrowers may close or default on their loan. This can be interpreted as an option. The lender sells the property in order to recover outstanding debt. The sale of the property is on the lender's terms and true market value may not be obtained. In a market where the differential—property value minus mortgage value—decreases with time or goes negative, the likelihood of exercising the option would increase, in order to reduce the borrower's losses. A lender would ordinarily be aware of the difference between mortgage value and property value, and current levels of interest rates, but non-mortgage information would not be known to the lender and hence whether the borrower is likely to exercise the option. For the property value greater than the mortgage value:

Net cash flow at T, $Y_{1T} - Y_{2T}$: Property value minus mortgage value minus transaction cost; $Y_{1i}, Y_{2i} = 0, i = T + 1, T + 2, \ldots, n$; there may also be a reputation (credit rating) cost extending beyond T.

Change

Changing the nature of a mortgage may be possible at some time in the future. For example, it may be possible to change between loans with different interest types—fixed rate (constant over a defined period or term) and variable (floating, adjustable) rate. This works like refinancing. The interest rate in variable rate loans generally fluctuates with the cash rate, which is set by a central bank, and fluctuates over time. When the cash rate is adjusted, commercial lenders make a decision on whether to pass on the same adjustment to their borrowers fully, partially or not at all. Commercial lenders may also adjust their rates independently of any cash rate movement. Borrowers and their loan repayments are thus exposed to the uncertainty of interest rate movements. Variable rate loan repayments increase with a rise in interest rates and decrease with a fall in interest rates. By comparison, fixed interest rate loans allow borrowers to lock in a definite interest rate for a given period or term; borrowers know exactly what their loan repayments are over the life of the loan. Fixing the interest rate protects the borrower from cash rate rises but not against rate falls, while such loans tend to be less flexible than variable rate loans, with restrictions on extra repayments and access to these, and also perhaps incurring greater costs for early termination. Having the ability to convert or switch between rate types can be viewed as an option.

Net cash flow at T, $Y_{1T} - Y_{2T}$: present worth (discounted to time T) of remaining (years > T) mortgage payments required before converting, minus the present worth (discounted to time T) of remaining (years > T) mortgage payments required after converting, minus transaction cost.

Repayments greater than required by the lender nay also be allowed. This will have the effect of reducing the principal and future interest amounts. Although not conventionally seen as an option, its value can be obtained similarly to the case just discussed.

12.5.7 Leases

Options can exist within leases. For example, Oppenheimer [6] mentions subleasing as a call option held by the lessor, and cancellation as a put option held by the lessee.

Others mention lease renewals as call options held by the lessor with renewal terms indexed to, for example, inflation or sales; the option to purchase at the end of a lease period is a call option held by the lessor; the case where the lessee receives a share of sales above a threshold is a call option held by the lessee.

The evaluation of all these options follows the same path as previously treated options. Estimates, including uncertainty, are made for equivalent cash inflows and cash outflows at time T with respect to the person holding the option; cash flows related to the option exercising but extending beyond time T are discounted to time T. Collectively, all the cash flows are discounted to time 0 and the option value calculated from Eq. (1.1).

12.6 Conclusion

Where the future movements in underlying prices or values or proxies are anticipated to be the same as the past, Black-Scholes, binomial lattices and equivalent Monte Carlo simulation, albeit with modification, can be used to value options. There is a large literature on this. Assumptions may be needed to extend such financial options tools to real options. Where the market is atypical, such methods may not be appropriate. This book's approach, however, has relaxed assumptions and can deal, for example, with any market irregularities or regularities as well as incorporate uncertainty in all the analysis variables.

The book's approach gives option values close to that of Black-Scholes. Trends in behaviour with rspect to analysis variables are also the same.

Property and index prices are uncertain, and real estate options provide one way of dealing with the associated risk. In atypical markets, including those influenced by political, economic and external factors, traditional option pricing tools may not apply. This book gives a practical method, suitable in such atypical markets as well as regular markets, by which real estate options could be valued. The approach is based on recognizing cash flows or cash flow equivalents that result from exercising an option, and the using a probabilistic present worth analysis. The approach's strengths are that it is intuitive to understand, straightforward to implement and requires low mathematical sophistication. Estimates of the cash flows or cash flow equivalents, including their uncertainty, can be made by any means available, and can incorporate experience, knowledge of the market, knowledge of anticipated shifts in economies or external factors, available time series or trends, or any other information known. There is no prescribed underlying price time series to follow, or method by which estimates should be obtained. The estimating method and incorporated knowledge is discretionary; all that is wanted is the best estimates possible.

It is anticipated, as with Black-Scholes, that users of the book's approach will adjust the approach over time to suit their experiences and particular requirements.

The chapter explored different real estate options including those related to property, indices, mortgages and leases, and demonstrated how these might be valued in atypical markets.

References

1. ASX (2008) *Introduction to index futures and options*. Australian Stock Exchange, Sydney, viewed 22 March 2018. https://www.asx.com.au/documents/products/intro_to_index_futures_and_options.pdf
2. ASX (2018) *Features*. Australian Stock Exchange, Sydney, viewed 22 March 2018. https://www.asx.com.au/products/index-options.htm
3. Black F, Scholes M (1973) The pricing of options and corporate liabilities. J Polit Econ 81(3):637–654
4. Carmichael DG (2014) Infrastructure Investment: an engineering perspective. CRC Press, Taylor and Francis, London
5. Carmichael DG (2016) A cash flow view of real options. Eng Econ 61(4):265–288
6. Oppenheimer PH (2002) A critique of using real options pricing models in valuing real estate projects and contracts. BriefS R Estate Financ 2(3):221–233

Selective Bibliography

The numerous author references contained in the books marked * are omitted from the following list. For omissions: https://www.engineering.unsw.edu.au/civil-engineering/staff/david-carmichael.

Investment

1. Carmichael DG (2017) Adjustments within discount rates to cater for uncertainty—guidelines. Eng Econom 62:322–335
2. Carmichael DG (2016) A cash flow view of real options. Eng Econom 61:265–288
3. Carmichael DG, Lea KA, Balatbat MCA (2016) the financial additionality and viability of CDM projects allowing for uncertainty. Environ Dev Sustain 18:129–141
4. Carmichael DG, Ballouz JJ, Balatbat MCA (2015) Improving the attractiveness of CDM projects through allowing and incorporating options. Energy Policy 86:784–791
5. Carmichael DG, Handford LB (2015) A note on equivalent fixed-rate and variable-rate loans; borrower's perspective. Eng Econom 60(2):155–162
6. *Carmichael DG (2014) Infrastructure investment: an engineering perspective. CRC Press, Taylor and Francis, London, UK
7. Carmichael DG, Bustamante BL (2014) Interest rate uncertainty and investment value—a second order moment approach. Int J Eng Manag Econom 4(2):176–189
8. Carmichael DG, Balatbat MCA (2011) On the analysis of property unit sales over time. Int J Strateg Prop Manag 15:329–339
9. Carmichael DG, Hersh AM, Parasu P (2011) Real options estimate using probabilistic present worth analysis. Eng Econom 56:295–320
10. Carmichael DG (2011) An alternative approach to capital investment appraisal. Eng Econom 56:123–139
11. Carmichael DG, Balatbat MCA (2008) Probabilistic DCF analysis and capital budgeting and investment—a survey. Eng Econom 53(1):84–102
12. Carmichael DG, Balatbat MCA (2008) The influence of extra projects on overall investment feasibility. J Financ Manag Prop Constr 13(3):161–175

D. G. Carmichael, *Future-proofingâ Valuing Adaptability, Flexibility, Convertibility and Options*, Management in the Built Environment,
https://doi.org/10.1007/978-981-15-0723-6

Adaptability, Flexibility and Convertibility

13. Carmichael DG (2018) Estimating the value of built-in flexibility in infrastructure. In: Wang C, Harper C, Lee Y, Harris R, Berryman C (eds) ASCE construction research congress 2018: sustainable design and construction and education, ASCE, New York, pp. 613–622, ISBN (PDF) 9780784481301, ASCE Construction Research Congress, New Orleans, Louisiana, April 2–5, 2018
14. Carmichael DG (2018) The case for building in adaptability in houses, joint asia-pacific network for housing research (APNHR) and Australasian housing researchers (AHR) conference, Griffith University Cities Research Institute, Gold Coast, 6–8 June 2018
15. Carmichael DG, Taheriattar R (2018) Valuing deliberate built-in flexibility in houses—exampled. Int J Strateg Prop Manag 22:479–488
16. Carmichael DG (2015) Incorporating resilience through adaptability and flexibility. Civil Eng Environ Syst 32:31–43
17. Carmichael DG, Hersh AM, Parasu P (2011) Real options estimate using probabilistic present worth analysis. Eng Econom 56:295–320

Risk

18. Carmichael DG (2016) Risk—a commentary. Civil Eng Environ Syst 33:177–198
19. Carmichael DG, Edmondson CG (2015) Risk in stream and royalty financing of infrastructure development. J Infrastruct Develop 1:23–40
20. Carmichael DG, Balatbat MCA (2011) Risk associated with managed investment primary production projects. Int J Proj Organ Manag 3:273–289

Problem Solving

21. Carmichael DG (2015) Incorporating resilience through adaptability and flexibility. Civil Eng Environ Syst 32:31–43
22. *Carmichael DG (2013) Problem solving for engineers. CRC Press, Taylor and Francis, London, UK
23. Carmichael DG (2013) The conceptual power of control systems theory in engineering practice. Civil Eng Environ Syst 30:231–242

Planning

24. Carmichael DG (2009) Comments on delay analysis methods in resolving construction claims. Int J Constr Manag 9:1–12
25. Carmichael DG (2007) An alternative to earned value reporting on project performance. In: CRIOCM 2007 international research symposium on advancement of construction management and real estate, Sydney, presented at CRIOCM 2007 international research symposium on advancement of construction management and real estate, Sydney, 8–13 August 2007
26. *Carmichael DG (2006) Project planning, and control. Taylor & Francis, London, UK
27. Carmichael DG (1997) Re-engineering and work study. J Proj Constr Manag 3:95–107

28. Al-Sadek O, Carmichael DG (1992) On simulation in planning networks. Civil Eng Environ Syst, 59–68
29. *Carmichael DG (1989), Construction engineering networks. Ellis Horwood Ltd (Wiley), Chichester, UK

Management and Project Management

30. Carmichael DG (2018) Organisations as systems—difficulties in model development and validation. Civil Eng Environ Syst 35(1–4):41–56
31. *Carmichael DG (2004) Project management framework. A. A. Balkema, Rotterdam, The Netherlands
32. Carmichael DG (1998) Gurus of Faddish management. J Proj Constr Manag 4:77–84
33. Carmichael DG (1997) Management fads. J Proj Constr Manag 3:115–125
34. Carmichael DG (1996) Flat organisational structures. J Proj Constr Manag 2:61–68
35. Carmichael DG (1996) On project objectives and constraints. J Proj Constr Manag 2:63–71

Construction, Quarrying and Surface Mining Emissions

36. Carmichael DG, Shen X, Peansupap V (2019) The relationship between heavy equipment cost efficiency and cleaner production in construction. J Clean Prod 211:521–529
37. Carmichael DG, Mustaffa NK (2018) Emissions and production penalties/bonuses associated with non-standard earthmoving loading policies. Constr Innov 18(2):227–245
38. Carmichael DG, Mustaffa NK, Shen X (2018) A utility measure of attitudes to lower-emissions production in construction. J Clean Prod 202:23–32
39. Jukic D, Carmichael DG (2016) Emission and cost effects of training for construction equipment operators: a field study. Smart Sustain Built Environ 5:96–110
40. Kaboli AS, Carmichael DG (2014) Optimum scraper load time and fleet size for minimum emissions. Int J Constr Manag 14:209–226
41. Kaboli AS, Carmichael DG (2014) Truck dispatching and minimum emissions earthmoving. Smart Sustain Built Environ 3:170–186
42. Carmichael DG, Bartlett BJ, Kaboli AS (2014) Surface mining operations: coincident unit cost and emissions. Int J Min Reclam Environ 28:47–65
43. Carmichael DG, Lea LO, Balatbat MCA (2014) Emissions management of urban earthmoving fleets. In: Tamosaitiene J, Panuwatwanich K, Mishima N (eds) Fifth international conference on engineering, project, and production management. Nelson Mandela Metropolitan University, Nelson Mandela Metropolitan University, Port Elizabeth, South Africa, pp 262–271, 26–28 November 2014
44. Carmichael DG, Malcolm CJ, Balatbat MCA (2014) Carbon abatement and its cost in construction activities. In: Construction research congress 2014: construction in a global network—proceedings of the 2014 construction research congress, pp 534–543
45. Carmichael DG, Williams EH, Kaboli AS (2012) Minimum operational emissions in earthmoving. In: Construction research congress 2012, proceedings of the 2012 ASCE construction research congress, pp 1869–1878

Contractor Payments

46. Tran H, Carmichael DG (2013) A contractor's classification of owner payment practices. Eng Constr Archit Manag 20:29–45
47. Tran H, Carmichael DG (2012) The likelihood of subcontractor payment: downstream progression via the owner and contractor. J Financ Manag Prop Constr 17:135–152

48. Tran H, Carmichael DG (2012) Contractor's Financial Estimation Based on Owner Payment Histories. Org Technol Manag Constr Int J 4:481–489
49. Carmichael DG, Balatbat MCA (2010) A contractor's analysis of the likelihood of payment of claims. J Financ Manag Prop Constr 15:102–117

Contracts and Project Delivery

50. Carmichael DG, Nguyen TA, Shen X (2019) Single treatment of PPP road project options. ASCE J Constr Eng Manag 145(2):04018122-1–04018122-11
51. Carmichael DG, Karantonis JP (2015) Construction contracts with conversion capability: a way forward. J Financ Manag Prop Constr 20:132–146
52. Carmichael DG (2009) Comments on delay analysis methods in resolving construction claims. Int J Constr Manag 9:1–12
53. Carmichael DG (2007) Delivery methods and their risk profiles, CRIOCM 2007 international research symposium on advancement of construction management and real estate, Sydney, presented at CRIOCM 2007 international research symposium on advancement of construction management and real estate, Sydney, 8–13 August 2007
54. *Carmichael DG (2002) Disputes and international projects. A. A. Balkema Publishers, Rotterdam, The Netherlands
55. *Carmichael DG (2000) Contracts and international project management. A. A. Balkema, Rotterdam, The Netherlands

Construction, Quarrying and Surface Mining Operations: Production and Cost

56. Carmichael DG (1991) Approximations for Heterogeneous Finite Source Queues. Civil Eng Syst 8:174–178
57. Carmichael DG (1990) Heterogeneity in deterministic finite source queues. Civil Eng Syst 7:11–19
58. Carmichael DG (1989) Production Tables for Earthmoving, Quarrying and Open-cut Mining. In: Carmichael DG (ed) Applied construction management. Unisearch, The University of New South Wales, Sydney, pp 275–284
59. Carmichael DG (1988) Queue simulation of cyclic construction operations. Civil Eng Syst 5:213–219
60. *Carmichael DG (1987) Engineering queues in construction and mining. Ellis Horwood Ltd (Wiley), Chichester, UK
61. Carmichael DG (1987) On the equivalence of the (Eh/M/c) and (M/M/c) finite source queues. Civil Eng Syst 4:87–93
62. Carmichael DG (1987) Optimal pusher-scraper loading policies. Eng Optim 12, 255–267
63. Carmichael DG (1987) Machine interference with general repair and running times. Zeitschrift fur Oper Res 31:B115–B133
64. Carmichael DG (1987) A refined queuing model for earthmoving operations. Civil Eng Environ Syst 4:153–159
65. Carmichael DG (1986) Shovel-truck queues: a reconciliation of theory and practice. Constr Manag Econom 4:161–177
66. Carmichael DG (1986) Optimal Shovel-truck operations. Eng Optim 10:51–63
67. Carmichael DG (1986) Erlang loading models in earthmoving. Civil Eng Syst 3:118–124
68. Carmichael DG (1986) The inclusion of storage in a queueing system. Zeitschrift für Oper Res 30:B127–B134

Optimum Structural Design

69. Carmichael DG (1991) Two notes on structural optimization. In: Franchi A, Riva P, Grierson D (eds) Progress in structural engineering. M. Z. Cohn Anniversary Volume, Kluwer, Dordrecht
70. Carmichael DG (1990) Bang-bang control and optimum structural design. Eng Optim 15:205–209
71. Carmichael DG (1990) Structural optimization and system dynamics. Struct Optim 2:105–108
72. *Carmichael DG (1981) Structural modelling and optimization. Ellis Horwood Ltd (Wiley), Chichester, UK
73. Carmichael DG (1980) Computation of pareto optima in structural design. Int J Numer Method Eng 15:925–929
74. Carmichael DG, Clyde DH (1980) On the optimum flexure problem in structural mechanics. Optim Control Appl Method 1:143–154
75. Carmichael DG, Clyde DH (1980) Multilevel control concepts in relation to the structural design problem. In: Leipholz HHE (ed) Structural control. North Holland, Amsterdam, pp 171–198
76. Carmichael DG (1979) Lyapunov theory in structural design. Eng Optim 4:159–163
77. Carmichael DG (1978) Probabilistic design of structures in state equation form. Eng Optim 3:83–92
78. Carmichael DG (1978) Optimal control in the design of material continua. Arch Mech 30:743–755
79. Carmichael DG (1977) On a minimum weight disk design problem. J Appl Mech Trans ASME 44:506–507
80. Carmichael DG (1977) Optimal vibrating plates and a distributed parameter singular control problem. Int J Control 26:19–31
81. Carmichael DG (1977) Singular optimal control problems in the design of vibrating structures. J Sound Vib 53:245–253
82. Carmichael DG (1977) On a minimum weight disk problem. J Appl Mech 44:506–507
83. Carmichael DG (1977) Singular optimal control problems in the design of vibrating structures'. J Sound Vib 53:245–253

Material and Structure Characterisation

84. Carmichael DG (1982) Adaptive filtering in structural dynamics. Eng Optim 5:235–247
85. Carmichael DG (1980) Identification of cyclic material constitutive relationships. In: Proceedings seventh Australasian conference on the mechanics of structures and materials. University of Western Australia, Perth, University of Western Australia, Perth, pp 130–132, 12–14 May
86. Carmichael DG (1979) The state estimation problem in experimental structural mechanics, third international conference—applications of statistics and probability to soil and structural engineering. The University of New South Wales, Sydney, The University of New South Wales, Sydney, pp 802–815, 29 January–2 February 1979
87. Carmichael DG (1979) Optimal filtering of concrete creep data, seventh canadian congress on applied mechanics. University of Sherbrooke, Quebec, University of Sherbrooke, Quebec, pp 121–122, 27 May–1 June 1979

Index

D. G. Carmichael, *Future-proofingâ Valuing Adaptability, Flexibility, Convertibility and Options*, Management in the Built Environment,
https://doi.org/10.1007/978-981-15-0723-6

MIX
Papier aus verantwortungsvollen Quellen
Paper from responsible sources
FSC® C105338

If you have any concerns about our products,
you can contact us on
ProductSafety@springernature.com

In case Publisher is established outside the EU,
the EU authorized representative is:
Springer Nature Customer Service Center GmbH
Europaplatz 3, 69115 Heidelberg, Germany

Printed by Libri Plureos GmbH
in Hamburg, Germany